electric circuit analysis

$S =$ Sum

$N =$ Number Terms

$a = 1^{\underline{st}}$ "

$L =$ Last "

$$S = \frac{N}{2}(a + l)$$

$$S = \frac{100}{2}(1 + 100)$$

$$S = 50(101)$$

$$S = 5050$$

electric circuit analysis

BENJAMIN ZEINES

Instructor, RCA Institutes, Inc.

reston publishing company, inc., reston, virginia

To my wife, Bella,
and to Francine and Sandy
May for their patience
and understanding.

10 9 8 7 6 5 4 3 2 1

ISBN: 0–87909–244–0

Library of Congress Catalog Card Number: 78–187525
Printed in the United States of America

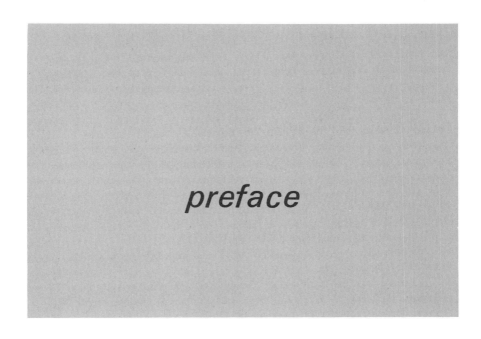

preface

This text introduces the basic concepts and mathematical techniques used in to-day's electrical and electronic industry. Educational institutions in the electrical technology and electrical engineering fields usually present a full year's study of both dc and ac circuit theory. This book is divided into two parts. The first half of this book presents the essential principles of dc circuit theory. The second half deals with the essential principles of ac circuit theory. Due to this writing technique, the text can readily be utilized for a two-semester course on circuit theory and analysis.

The theory is presented in as simple a manner as possible and, after the theory is presented, an illustrative problem is provided to demonstrate the technique. This procedure is followed throughout the text.

The text material should provide the student with an adequate background to enter into subsequent advanced courses of study either in power or in the electronics field.

The level of presentation assumes that the student has completed the required high school courses both in physics and mathematics and is currently studying technical mathematics in conjunction with the study of circuits.

Chapter 1 deals with units, dimensions, powers of ten and circuit definitions. Chapter 2 introduces the concept of electrical resistance both wire and semiconductor types. Chapter 3 initiates the relationship between voltage, current and power. Chapter 4 introduces dc circuits. This chapter is important in that it presents the idea that a single resistor can represent a number of series resistors or a number of parallel resistors. This concept is further utilized to present both dc meters and network theorems in Chapters 5 and 6 respectively. The use of determinants as a tool in the solution of network equations is demonstrated in Chapter 7. A circuit property of an electrical element called capacitance is presented in Chapter 8. Chapter 9 is an introduction to inductors which is presented in Chapter 10. This comprises the first half or dc half of the text.

Chapter 11 introduces the method used to produce alternating current. Chapter 12 treats with the mathematics necessary to handle complex quantities. Chapters 13 and 14 applies the complex number theory to both series and parallel circuits respectively. Chapter 15 introduces the concept of resonance. Chapter 16 presents network theorems and analysis techniques for use in ac circuits. Chapter 17 deals with the problem of the transformer and transformer circuits. The problem of non-sinusoidal waveforms is treated in Chapter 18, both analytically and graphically. For students who have difficulty with the mathematics involved, the graphical technique should prove adequate.

Additional problems for student assignment and testing are included within and at the end of each chapter.

I would like to acknowledge the constructive criticism given me by my colleagues Mr. Jerry Wertzer and Mr. Mark Pallans. I wish to thank Mrs. A. Thalrose for her typing of the manuscript. Special thanks go to Mrs. Bella Zeines for her patience, forbearance, and continuous encouragement during the time this manuscript was in preparation.

BEN ZEINES

contents

1 *introduction*   1

units, 1   powers of ten, 2   conversion units, 4
properties of charged bodies, 5   current, 6
potential difference, 8   conductors and insulators, 10
problems, 10

2 *resistance*   13

introduction, 13   circular mil, 15   wire tables, 16
temperature effects, 18   temperature coefficients, 19
color code for commercial resistors, 20
semiconductor resistors, 22   electrons and holes, 23
p-type silicon, 24   n-type silicon, 25
diffused silicon resistors, 26   **problems**, 27

3 *current, voltage, and power*   29

circuit symbols, 29   Ohm's law, 30   power, 31
efficiency, 33   energy, 34   **problems**, 35

4 *dc circuits*   37

series circuits, 37   voltage dividers, 42   parallel circuits, 46
current dividers, 49   series parallel circuits, 51
problems, 54

5   dc meters   59

introduction, 59   D'Arsonval meter, 59   ammeters, 61
voltmeters, 64   voltage sensitivity, 68   ohmmeters, 70
Wheatstone bridge, 72   **problems**, 74

6   network theorems   76

π to T conversion, 76   T to π conversion, 79
superposition theorem, 82   Thevenin's theorem, 88
Norton's theorem, 91   Millman's theorem, 92
reciprocity theorem, 93   maximum power transfer theorem, 95
loop circuit analysis, 96   nodal analysis, 99
problems, 102

7   determinants   108

introduction, 108   determinant of order n, 113
problems, 115

8   capacitance   118

introduction, 118   electric field, 118   flux density, 119
capacitance, 120   capacitors in series and parallel, 123
capacitors in parallel, 125   types of capacitors, 126
RC charging circuit, 129   time constant, 132
RC discharge circuit, 134   energy stored in a capacitor, 137
problems, 138

9   magnetic circuits   142

introduction, 142   magnetic fields, 143
magnetomotive force, 144   reluctance, 144
permeability, 145   flux density and magnetizing force, 145
magnetism, 146   hysteresis curves, 147
magnetic circuits, 149   air gaps, 155
parallel magnetic circuits, 157   **problems**, 159

10   inductors   164

introduction, 164   Faraday's law, 164   Lenz's law, 166
self-induction, 166   inductances in series and parallel, 168
RL circuit, 170   energy stored by an inductance, 174
RL discharging circuits, 175   **problems**, 179

11   alternating current   183

rotating generator, 183   the sine wave, 185
phase angle, 187   alternating currents, 190
average value, 196   power and power factor, 197
ac circuits with resistance, 199
ac circuit with inductance, 200
ac applied to a pure C circuit, 202   **problems**, 206

12   *phasor algebra*   **211**

complex quantities, 211   *conversion methods*, 215
mathematical operations, 218   *phasors*, 226
problems, 229

13   *ac circuits*   **232**

introduction, 232   *pure R circuit*, 232
pure L circuit, 235   *pure C circuit*, 237
series RL circuit, 238   *series RC circuit*, 244
series circuits containing more than two elements, 249
series RLC circuit, 251   ***problems***, 254

14   *parallel circuits*   **258**

resistors in parallel, 258   *parallel RC circuit*, 259
conductance, susceptance, and admittance, 262
parallel RL circuit, 265   *impedance and admittance*, 269
parallel RLC circuits, 276   *series parallel ac circuits*, 281
problems, 283

15   *resonance*   **288**

series resonant circuit, 288   *sharpness of resonance*, 291
maximum voltage considerations, 293   *parallel resonance*, 295
parallel resonance with resistance in both branches, 298
sharpness of resonance, 301   *maximum impedance*, 304
multiple resonance, 305   ***problems***, 307

16   *analysis techniques*   **311**

introduction, 311   *loop equations*, 311   *nodal analysis*, 314
conversion of voltage and current sources, 317
superposition theorem, 318   *Thevenin's theorem*, 320
Norton's theorem, 322   *maximum power transfer theorem*, 323
T to π and π to T conversions, 325
equivalent T representation of complex network, 328
four-terminal network parameters, 331   ***problems***, 335

17   *transformers*   **341**

mutual induction, 341   *transformation ratio*, 345
polarity of coils, 347   *transformer analysis*, 348
transformer equivalent circuits, 350   *audio transformer*, 354
transformer testing, 355   ***problems***, 358

18   *nonsinusoidal waveform analysis*   **362**

Fourier series, 362   *symmetry*, 366
graphical techniques, 373   *power considerations*, 377
problems, 378

A   *slide rule instructions*   385

B   *table of exponentials*   405

C   *natural trigonometric functions*   409

        index   417

1

introduction

units

Inherent in the analysis of electrical circuits is a requirement that the units of electrical quantities must be either described or calculated. The basic electrical quantities are defined in the units of either the English or the metric system. The English system of units is based on the length, mass or weight, and second system. The metric system comprises two sections: the MKS system and the CGS system. The MKS system is based on the meter, kilogram, and second system, whereas the CGS system is based on the centimeter, gram, and second system.

Most textbooks in electrical circuit theory use the MKS system of units. Consequently, this text will adhere to this particular system. The three basic units common to all systems are the mass (M), length (L), and time (sec) units. Table 1–1 compares the three systems of units.

In the MKS system, *the meter* was originally defined as equal to a distance specified by two marks spaced a certain distance apart on a platinum-iridium bar. This bar is maintained at the International Bureau of Weights and Measures at Sevres, France.[1]

One kilogram was defined as equal to the mass of water at 4°C in a cube 1000 cubic centimeters in volume.[2]

---

[1]The meter was recently defined (in 1960) as 1,650,763.73 wavelengths of krypton 86, orange line.

[2]The modern standard defines the kilogram as the mass of a particular platinum-iridium cylinder and the liter as the volume occupied by 1 kg of water at 4°C.

table 1–1

unit	English	MKS	CGS
length	1 yard	1 meter	1 centimeter $= 0.01$ m
	1 foot $= \frac{1}{3}$ yd	100 cm $= 1$ m	
	1 inch $= \frac{1}{12}$ ft		
mass	1 slug $= 14.6$ kg	1 kg	1 gram $= 0.001$ kg
weight	1 pound $= 4.45$	1 N	1 dyne
	newtons (N)	10^5 dyn $= 1$ N	
time	second (sec)	second	second

conversion units	
1 m	39.37 in.
2.54 cm	1 in.
1000 grams $= 1$ kg	2.205 lb
1000 liters $= 1$ kiloliter	
1000 milliliter $= 1$ liter	1000 cubic centimeters $= 1.057$ quarts

One second was defined as the ratio of

$$\frac{\text{mean solar day}}{86,400}$$

The modern definition of the second is

$$\frac{\text{the tropical year 1900}}{31,556,925.9747}$$

powers of ten

In electrical circuit theory, one is often required to operate with extremely large or extremely small numbers. This problem can be solved most conveniently by using powers of ten. Some examples of this notation are

$$10,000 = 10^4$$
$$12,000,000 = 12 \times 10^6$$
$$0.00001 = 10^{-5}$$

Many of these large or small numbers appear quite often in electrical applications. Consequently, standard forms of abbreviations were adopted as listed in Table 1–2.

table 1–2

power of 10	name (prefix)	abbreviation
10^{12}	Tera	T
10^9	Giga	G
10^6	Mega	M
10^3	Kilo	k or K
10^{-3}	milli	m
10^{-6}	micro	μ
10^{-9}	nano	n
10^{-12}	pico	p

A few examples of the forms in Table 1–2 are

$$12{,}000{,}000 \text{ ohms} = 12 \times 10^6 \text{ ohms} = 12 \text{ megohms}$$
$$3000 \text{ meters} = 3 \times 10^3 \text{ meters} = 3 \text{ kilometers}$$
$$0.015 \text{ sec} = 15 \times 10^{-3} \text{ sec} = 15 \text{ millisec}$$

Consider a few examples demonstrating the use of powers of ten and the various algebraic operations.

(a)
$$5800 + 74{,}200 = 5.8 \times 10^3 + 74.2 \times 10^3$$
$$= (5.8 + 74.2)\ 10^3$$
$$= 80 \times 10^3$$

(b)
$$768{,}400 - 58{,}400 = 76.84 \times 10^4 - 5.84 \times 10^4$$
$$= (76.84 - 5.84)\ 10^4$$
$$= 71 \times 10^4$$

(c)
$$(0.00055)(0.000003) = (55 \times 10^{-5})(3 \times 10^{-6})$$
$$= 165 \times 10^{-11}$$

(d)
$$(480{,}000)(0.0000125) = 4.8 \times 10^5 \times 125 \times 10^{-7}$$
$$= 6$$

(e)
$$\frac{0.000048}{0.002} = \frac{48 \times 10^{-6}}{2 \times 10^{-3}} = 24 \times 10^{-3}$$

(f)
$$\frac{780{,}000}{0.00000013} = \frac{78 \times 10^4}{13 \times 10^{-8}} = 6 \times 10^{12}$$

(g) $(0.00006)^2 = (6 \times 10^{-5})^2 = 36 \times 10^{-10}$

(h) $(500{,}000)^3 = (5 \times 10^5)^3 = 125 \times 10^{15}$

conversion units

In some cases, it may be desirable to convert from one system of units to another. Consider the process of converting 100,000 grams to kilograms. Thus,

$$\frac{1 \text{ kg}}{1000 \text{ g}} = 1 \quad \text{or} \quad \frac{1000 \text{ g}}{1 \text{ kg}} = 1$$

It is evident that since our result must be in kilograms, the multiplying factor must be

$$\frac{1 \text{ kg}}{1000 \text{ g}}$$

Thus,

$$100,000 \text{ g} \times \frac{1 \text{ kg}}{1000 \text{ g}} = 100 \text{ kg}$$

Consider the problem of converting 100,000 kg to grams. Thus,

$$100,000 \text{ kg} \times \frac{1000 \text{ g}}{1 \text{ kg}} = 10^5 \times 10^3 \text{ g}$$
$$= 10^8 \text{ g}$$

Convert 0.00000000156 sec to nanoseconds.

$$156 \times 10^{-11} \text{ sec} \times \frac{1 \text{ nsec}}{10^{-9} \text{ sec}} = 1.56 \text{ nsec}$$

In other cases, it may be necessary to convert from one system of units to another. Some sample problems will demonstrate the conversion technique.

Convert 76,300 meters to yards.

$$76,300 \text{ m} \times \frac{39.37 \text{ in.}}{1 \text{ m}} \times \frac{1 \text{ yd}}{36 \text{ in.}} = 8.346 \times 10^4 \text{ yd}$$

Convert 68 lb to newtons.

$$68 \text{ lb} \times \frac{4.45 \text{ N}}{1 \text{ lb}} = 302.6 \text{ N}$$

Convert 120,000 ft to millimeters.

$$120,000 \text{ ft} \times \frac{12 \text{ in.}}{1 \text{ ft}} \times \frac{2.54 \text{ cm}}{1 \text{ in.}} \times \frac{10 \text{ mm}}{1 \text{ cm}} = 365.8 \times 10^5 \text{ mm}$$

Convert 4.8 hr to milliseconds.

$$4.8 \text{ hr} \times \frac{60 \text{ min}}{1 \text{ hr}} \times \frac{60 \text{ sec}}{1 \text{ min}} \times \frac{1000 \text{ msec}}{1 \text{ sec}} = 1.728 \times 10^8 \text{ msec}$$

properties of charged bodies

The study of electrical circuit theory requires a basic understanding of the atomic structure of matter. The structure of the atom is assumed to be similar to that of the solar system with the nucleus of the atom being analogous to the sun and the rotation of the electrons about the nucleus being analogous to the rotation of the planets about the sun. The rotation of the planets is considered coplanar, whereas the orbital motions of the electrons follow concentric three-dimensional spherical paths.

Each electron orbiting about the nucleus carries a negative charge equal to 1.602×10^{-19} coulomb. A corresponding proton exists within the nucleus for each electron, each proton having a positive charge equal to the electron charge. The atomic structure of an element can be represented by a nucleus and a number of concentric shells, as shown in Fig. 1–1.

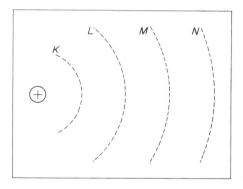

figure 1–1
atomic structure

The first shell closest to the nucleus is called the K shell and can accommodate only two electrons as a maximum. The second shell from the nucleus is called the L shell and accommodates a maximum of eight electrons. The third outer shell is called the M shell and will accept a maximum of 18 electrons. The fourth outer shell is called the N shell and will accept a maximum of 32 electrons. The maximum number of electrons that each shell can accept is determined by the equation $2n^2$, where n is the number of the shell as we count outward from the nucleus.

It has been determined experimentally that charges of like sign repel each other, whereas charges of unlike signs attract each other. The force of attraction or repulsion can be calculated by the formula

$$F = \frac{k\,Q_1 Q_2}{r^2} \text{ newtons} \qquad (1\text{-}1)$$

where

$$k = 8.998 \times 10^9 \text{ (constant)}.$$
$$Q_1, Q_2 = \text{charges in coulombs}.$$
$$r = \text{distance between the two charges in meters}.$$

It should be noted that the nucleus contains both protons (positive charges) and neutrons (neutral or zero charge). The atomic number of an element in the periodic table is defined by the total number of electrons in the various shells. The atomic weight of an element is defined by the total number of protons and neutrons within the nucleus. It is evident that the binding force maintaining the electrons in the inner shell must be greater than the force necessary to hold the electrons in an outer shell. Consequently, less energy must be expended to remove an electron from an outer shell.

In general, electrons can be readily removed when the outer shell is incomplete and possesses relatively few electrons. Consider the element copper, which contains 29 electrons. The atomic structure of copper is illustrated in Fig. 1–2.

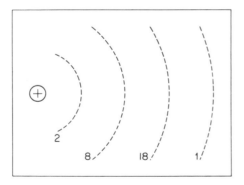

figure 1–2
atomic structure of copper

The application of sufficient energy to the element copper will cause the removal of the electron from the outer shell. The electron is then established as a *free electron*.

current

When atoms of copper are tightly packed together in the form of a length of wire, the single electron in the outer shell is free to wander from one atom to another. Thus, at room temperature, free electrons are in random motion throughout the wire. When there is no external force applied to the copper wire, the resultant flow of charge in any one direction is zero.

If the copper wire is connected to a battery, as shown in Fig. 1–3, free electrons within the copper wire will flow towards the positive (+) charge terminal. The *rate of flow* of these electrons constitutes a current. The unit

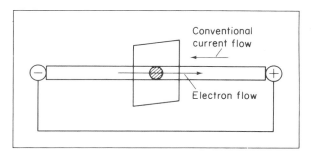

figure 1–3
electron flow in a conductor

of electric current is the ampere, abbreviated by the symbol A. Since the electron charge is specified by 1.602×10^{-19} coulombs, then it requires 6.242×10^{18} electrons per second to produce 1 ampere. Thus,

$$1 \text{ ampere} = (1.602)(6.242)10^{-1} \text{ coulombs/sec}$$

or 1 ampere is defined as equal to 1 coulomb per second passing through a specified region. The basic formula for the average current is

$$I = \frac{Q \text{ (coulombs)}}{t \text{ (second)}} \text{ amperes} \qquad (1\text{–}2)$$

It is evident that two directions for current flow exist. The current that flows from the positive terminal to the negative terminal is called *conventional current flow* and is used throughout this text. The flow of current from negative to positive is called *electron motion*.

It is evident that two possible directions of current flow exist. It would be convenient to establish one direction of current flow only in an electric circuit. Note that the definition of an ampere is determined by the magnitude of charges moved in a period of time, and is independent of direction. Consequently, since there are two possible directions, two different conventions have been adopted.

One convention is based on the negative test charge and is referred to as the *electron flow*. The second convention is based on the positive test charge and is referred to as the *conventional current flow*. The conventional current flow method is used extensively throughout the electrical engineering and technology fields.

Electrical current is then defined as the conventional current flow or positive charge carriers leaving the positive terminal of a source and entering the negative terminal of the source. Both conventions are illustrated in Fig. 1–4 (a) and (b) respectively.

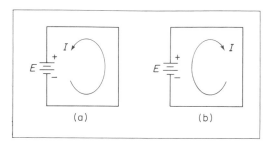

figure 1–4
(a) electron flow   (b) conventional current flow

potential difference

Potential difference may be considered as work per unit charge. The unit of work is the joule. One joule is equal to the work performed when a force of 1 newton is applied over a distance of 1 meter. Work must be performed to bring two positively charged bodies together. Consider the body shown in Fig. 1–4, which has lines of force radiating outwards and which has a positive charge of Q coulombs.

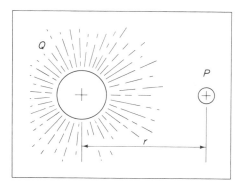

figure 1–5
effect of a large electric field on a point charge

A small charge of *p coulombs* is located at a distance *r* from the origin at *Q* coulombs. To move this small positive charge of *p* coulombs closer, work must be performed. Note that a potential difference will then exist between the two charges. The potential difference per unit test charge is determined by the equation

$$\text{Potential difference in volts} = \frac{W \text{ work in joules}}{Q \text{ charge in coulombs}} \tag{1-3}$$

One volt is defined as the potential difference existing when 1 joule of energy is involved in moving a charge of 1 coulomb from one point to another. An illustrative problem will demonstrate the theory.

sample problem

Ten joules of energy are applied to move a test charge of one coulomb from infinity to a value of *r* distance from the initial radiating lines of force. Determine the potential difference.

solution

1. Refer to the diagram illustrating the problem.

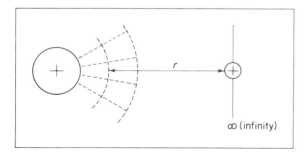

2. Calculate the potential difference.

$$\text{P.D. in volts} = \tfrac{10}{1} = 10 \text{ volts}$$

Conceptually, the difference of potential is the ability to perform work. The term *voltage* can refer either to a potential difference or to an electromotive force. There is a distinction between the two terms, although they are used interchangeably.

Consider a lamp plugged into an ac wall plug. The voltage at the wall plug is 110 volts. When the switch is closed, 110 volts is applied to the lamp and the lamp lights. When the switch is open, 110 volts exist at the wall plug, but the voltage across the lamp is zero. Consequently, the voltage at the wall plug is an emf, whereas the voltage applied to the lamp is considered a *potential difference* or a *voltage drop*.

conductors and insulators

Connection of a wire between the positive and negative terminals of a battery establishes a current flow through the wire. Different wires varying in size, type of material, etc., will produce different current flows within the wires. A *conductor* is a substance that offers little opposition or resistance to the flow of electrical currents. Metallic wires made of platinum, copper, or silver have extremely low electrical resistance and are referred to as good conductors.

An insulator is a substance that offers extremely high resistance to the flow of electrical current. For example, polystyrene is an excellent insulator, because it prevents almost any current flow. Consequently, it is used as an insulating material to cover current-carrying wires; it permits handling of these wires during current flow.

There is a class of materials that exists between conductors and insulators that is classified as semiconductors. These materials are used in the manufacture of semiconductor transistors.

Conductors that offer minimum resistance to electric current flow are symbolized by the letter R, and the unit of resistance is the *ohm*. One ohm may be defined as the resistance of a circuit element which permits a current of 1 ampere to flow through the circuit element when a steady voltage of 1 volt is applied across the circuit element. Thus,

$$1 \text{ ohm} = \frac{1 \text{ volt}}{1 \text{ ampere}}$$

$$R = \frac{E}{I} \text{ ohms}$$

problems

1. Express the following numbers as powers of ten.

(a) 0.000025

(b) 1,354,000,000

(c) 0.00000000012

(d) 1,642,000

(e) 0.00028

(f) 6750

(g) 0.002

(h) 8,340,000,000,000

2. Perform the following operations and express the resultant as powers of ten.
   (a) $10^3 (10^5)$
   (b) $10^6 (10^{-12})$
   (c) $10^{-3} (10^{-9})$
   (d) $10^6 (10,000,000,000)$
   (e) $10,000 (0.0000000001)$
   (f) $100,000 (10^{-9}) (10^{18})$
   (g) $\dfrac{1000}{0.000001}$
   (h) $10^{24} (10^{-30}) (10^{16})^{1/2}$

3. (a) $(100)^4$
   (b) $(0.000081)^{1/2}$
   (c) $(0.000000125)^{1/3}$
   (d) $(0.000064)^{1/2} (0.125)$
   (e) $\dfrac{(100)^3 (0.000000000005)}{0.00025}$
   (f) $\dfrac{(100)^{-5} (1000)^2}{(0.00025)}$

4. Convert the following:
   (a) 59 kg to grams
   (b) 65,000 m to kilometers
   (c) 0.0005 sec to milliseconds
   (d) 5650 mm to meters
   (e) 105,630 millisec to hours
   (f) 0.000000000015 sec to nanoseconds
   (g) 0.000058 g to micrograms

5. A uniform current of 2×10^{-16} amperes is flowing through a wire. Determine the number of electrons.

6. How many electrons flow through a wire in one hour if the current is 0.36 ampere?

7. The current flowing through a wire fuse is 3.00 amperes. Determine the length of time for 36 coulombs to pass through the fuse.

8. How many coulombs of electrons pass through a switch if the current is 300 milliamperes (ma) per minute?

9. Determine the length of time required for 30 coulombs to pass through a switch when the current is 240 ma.

10. Determine the potential difference if 25 joules of energy moves 5 coulombs of electrons between the two terminals.

11. What energy is necessary for 50 millicoulombs to pass through a potential difference of 200 volts?

12. In an amplifying device, 40 coulombs of charge is moved through a potential difference of 300 volts. How much work in joules is performed?

13. How many coulombs of charge will be moved when the applied energy is 0.5 joule and the potential difference is 1.5 volts?

14. A voltage of 5 volts produces a current of 200 ma through a conductor. Calculate the ohmic resistance of the conductor.

15. A 30-ampere current is flowing through a conductor having a resistance of 3 ohms. What is the voltage across the conductor?

2

resistance

introduction

The opposition that a conductor offers to the flow of current is called *resistance*. This opposition to the charge flow converts electrical energy to heat. This energy conversion is unidirectional. Heat created by the application of electrical energy in the resistor is radiated or conducted away. In some cases, this conversion into heat is considered desirable, as in electric heaters, electric blankets, electric irons, etc. In most electrical circuit applications, heat is an undesirable by-product in the transmission of energy and is kept at a minimum level.

The resistance of any wire conductor depends on the length, type of material, temperature, and cross-sectional area. The resistance of any conductor at a specified temperature (usually 20°C) can then be expressed mathematically as

$$R = \rho \frac{l}{A} \text{ ohms} \qquad (2\text{--}1)$$

where

l = length of the conductor in meters.
A = cross-sectional area in square meters.
ρ = resistivity in ohm meters.

The resistance is, therefore, directly proportional to the length and resistivity of the conductor and inversely proportional to the cross-sectional area of the conductor. It should be noted that the resistivity of a specified material is a constant and is independent of the shape, size, or structure of the material. The values of resistivity for some common materials are shown in Table 2–1.

table 2–1   resistance constants for various metals at 20°C

material	ρ at 20°C ohm-meter	ρ at 20°C ohm-cmil	α at 20°C per degree C
aluminum	2.83×10^{-8}	17.00	0.00391
brass	7.00×10^{-8}	42.00	0.002
copper	1.72×10^{-8}	10.37	0.00393
gold	2.44×10^{-8}	14.7	0.0034
iron	10.00×10^{-8}	60.00	0.0055
lead	22.00×10^{-8}	132.00	0.0043
nickel	7.8×10^{-8}	47.00	0.006
silver	1.47×10^{-8}	9.90	0.0038

An illustrative problem will demonstrate the theory and the use of the given equation.

sample problem

Determine the resistance of a 50-meter length of copper having a cross-sectional area of 5×10^{-5} square meters.

solution

step 1:   From Table 2–1, determine ρ.

$\rho = 1.72 \times 10^{-8}$ ohm-meter

step 2:   Calculate R.

$$R = \rho \frac{l}{A}$$

$$R = 1.72 \times 10^{-8} \frac{50}{5 \times 10^{-5}}$$

$$R = 17.2 \times 10^{-3} \text{ ohm}$$

circular mil

In the United States, the diameter of a wire is measured in units of a thousandth of an inch called a *mil*. Consider the typical wire shown in Fig. 2–1. Note that the conductor has a diameter of 1 mil.

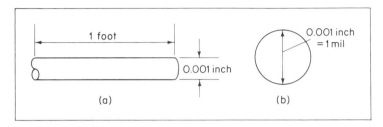

figure 2–1
wire constants: (a) mil foot; (b) cross section of a wire

To be consistent with the MKS system of units, the cross-sectional area of the conductor must be expressed in square meters. It is evident that the actual area of the circular wire is $\pi d^2/4$. This unit is impractical in dealing with the typical wire sizes used in electrical conductors. Resistance calculations can be simplified by eliminating the constant $\pi/4$. Consequently, a new unit of area applicable to wires or to conductors of circular cross section has been established. This new unit is called the *circular mil* and is defined as the cross-sectional area of a circle whose diameter is 1 mil. Mathematically, this relationship is

$$A = d^2 \text{ circular mils}$$

The abbreviation for circular mils is cmil.

A circular mil-foot (cmil-ft) is defined as the volume of a conductor that is one foot long and has a diameter of 1 mil. An example of this unit is shown in Fig. 2–1(a).

---

sample problem

Calculate the resistance of a copper wire 100 feet long having a diameter of 0.02 in.

---

solution

step 1:   From Table 2–1, the constants required are

$$\rho = 10.37 \ \frac{\text{ohm-cm}}{\text{foot}} \ ; \quad A = (20)^2 \text{ cmil}$$

step 2:   Calculate R.

$$R = \rho \frac{l}{A}$$

$$R = 10.37 \tfrac{100}{400} \text{ ohms}$$

$$R = 2.593 \text{ ohms}$$

---

sample problem

Determine the length of a copper wire having a diameter of 0.005 in. and a resistance of 10 ohms.

---

solution

step 1:   Calculate A in circular mils

$$A = (5)^2 \text{ cmil}$$

$$A = 25 \text{ cmil}$$

step 2:   The required formula is

$$l = \frac{RA}{\rho}$$

$$l = \frac{10 \times 25}{10.37} = 2.41 \text{ ft}$$

wire tables

Circular wire conductors are manufactured in sizes based on the American Wire Gauge (AWG) system. The gauge numbers define the size of wire in terms of its mil diameter and its cross-sectional area in circular mils. The wire with the largest diameter made has a gauge number of 0000. This number specifies a 460-mil diameter and a cmil area of 211,600. As the gauge numbers increase sequentially, the mil diameter and circular mil area become smaller, indicating a thinner wire. The American Wire Gauge system used for electrical conductors is tabulated in Table 2–2.

This table is based on the mathematical relationship $r = (D_1/D_2)^{1/n}$, where r is the common ratio of diameters of any two consecutive gauge numbers. D_1 is the larger diameter, and D_2 is the diameter of the smaller gauge number; n is the numerical difference between the two gauge numbers. An example will demonstrate the theory.

table 2-2   American wire gauge conductor sizes

AWG number	diameter in mils	area circular mils	AWG number	diameter in mils	area circular mils
0000	460.0	211,600	19	35.89	1288
000	409.6	167,800	20	31.96	1022
00	364.8	133,100	21	28.46	810.1
0	324.9	105,500	22	25.35	642.4
1	289.3	83,690	23	22.57	509.5
2	257.6	66,370	24	20.10	404.0
3	229.4	52,640	25	17.90	320.4
4	204.3	41,740	26	15.94	254.1
5	181.9	33,100	27	14.20	201.5
6	162.0	26,250	28	12.64	159.8
7	144.3	20,820	29	11.26	126.7
8	128.5	16,510	30	10.03	100.5
9	114.4	13,090	31	8.928	79.7
10	101.9	10,380	32	7.950	63.21
11	90.74	8234	33	7.080	50.13
12	80.81	6530	34	6.305	39.75
13	71.96	5178	35	5.615	31.52
14	64.08	4107	36	5.000	25.00
15	57.07	3257	37	4.453	19.83
16	50.82	2583	38	3.965	15.72
17	45.26	2048	39	3.531	12.47
18	40.30	1624	40	3.145	9.888

sample problem

Find the ratio of two gauge wires, numbers 26 and 20, respectively.

solution

step 1:   The value of n is 6. From Table 2-2,

$$D_1 = 31.962$$
$$D_2 = 15.941$$

step 2:   Calculate r.

$$r = \left(\frac{31.962}{15.941}\right)^{1/6}$$

$$r = 1.1225$$

sample problem

What is the resistance of 600 ft of no. 12 copper wire at 20°C?

solution

step 1:   From Tables 2–1 and 2–2, determine

$$\rho = 10.37; \qquad A = 6529.9 \text{ cmil}$$

step 2:   Calculate R.

$$R = \rho \frac{l}{A} = 10.37 \frac{600}{6529.9}$$

$$R = 0.952 \ \Omega$$

temperature effects

   The electrical resistance of most conductors varies with temperature. For the majority of metallic conductors the resistance increases with increasing temperature. The increase in resistance is due to the increased random motion of free electrons colliding more frequently with the charge carriers within the conductor. If resistance increases with temperature, the conductor is said to have a positive coefficient of temperature, and vice versa. Materials having a negative coefficient of temperature are the semiconductors, such as germanium, silicon, carbon, etc. The calculation of the resistance of a wire at various temperatures can be graphed as a straight line, as shown in Fig. 2–2.
   Experimental measurements of copper wire have established an absolute zero value of T for copper as $-234.5°C$. Using the method of similar triangles,

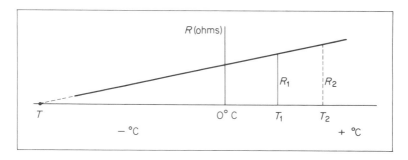

figure 2–2
variation of resistance with temperature

we find that a mathematical relationship exists whereby resistances at different temperatures may be calculated. Thus,

$$\frac{R_1}{234.5+T_1} = \frac{R_2}{234.5+T_2}$$

(2–2)

An illustrative problem will demonstrate the theory.

sample problem

A copper wire has a resistance of 24 ohms at 0°C. Determine the resistance at 35°C.

solution

$$\frac{24}{234.5+0} = \frac{R_2}{234.5+35}$$

$$R_2 = 24\frac{269.5}{234.5}$$

$$R_2 = 26.55 \ \Omega$$

The resistance of any copper conductor can readily be measured at the ambient temperature or room temperature. As energy is applied to the copper conductor, the temperature is increased. The resistance measurement under operating conditions will determine the amount of temperature increase.

temperature coefficients

Consider the mathematical relationship given by Equation (2–2). Thus,

$$\frac{R_2}{R_1} = \frac{234.5+T_2}{234.5+T_1}$$

This equation can be made general or valid for any element other than copper by converting the 234.5 back to T. Thus,

$$\frac{R_2}{R_1} = \frac{T+T_2}{T+T_1}$$

By some algebraic manipulation, we find that the resultant equation is

$$\frac{R_2}{R_1} = \frac{T+T_1+T_2-T_1}{T+T_1} = \frac{T+T_1}{T+T_1} + \frac{T_2-T_1}{T+T_1}$$

and

$$\boxed{R_2 = R_1\left[1+\alpha\left(T_2-T_1\right)\right]} \qquad (2\text{–}3)$$

where

$$\alpha = \frac{1}{T+T_1} \text{ (temperature coefficient of resistance).}$$

Note that if T_1 is equal to zero degrees centigrade, then R_1 is the resistance of the element at 0°C, and R_2 is the resistance of the element at the final temperature. The value of α is then determined at 0°C. Table 2–1 gives the value of α at 20°C for some common metallic elements.

An illustrative problem will demonstrate the theory.

---

sample problem

A copper wire has a resistance of 200 Ω at 20°C. Determine the resistance of the wire at 120°C.

---

solution

step 1: Determine the temperature coefficient of copper at 20°C.

From Table 2–1, $\alpha = 0.00393$.

step 2: Calculate the "hot" resistance.

$$R_{120°C} = R_{20°C}\left[1+\alpha\left(T_1-T_0\right)\right]$$
$$R_{120°C} = 200\left[1+0.00393(120-20)\right]$$
$$R_{120°C} = 278.6 \ \Omega$$

color code for commercial resistors

The materials used in the manufacture of commercial resistors are nichrome wire, forming wire-wound resistors, or a mixture of carbon with a binder that can be altered to form different values of resistance. The wire-wound resistor has nichrome wire wound in small helical coils, then dipped into a ceramic material and baked. This type of resistor is used in electric heaters, irons, broilers, etc. The carbon type of resistor is used where the current ratings are much smaller and consequently are manufactured in small sizes. It is physically much simpler to use a color code to denote the value of the resistor rather than to print the numerical value. The color code bands on a carbon resistor represent the order given in Table 2–3. Consider the diagram of a carbon resistor shown in Fig. 2–3.

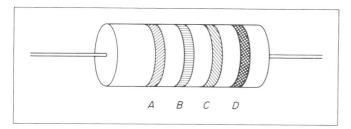

figure 2–3
color code

Four color bands are printed on one end of the resistor. Each color has the numerical value assigned, regardless of position. Bands A and B represent the first and second digits, respectively. Band C defines the number of zeros that follow the two digits; it is called the *multiplying factor*. The fourth band manufactures tolerance and defines the allowable deviation off the assigned value.

table 2–3   color code for carbon resistors

	A and B	C	D
black	0	10^0	
brown	1	10^1	
red	2	10^2	
orange	3	10^3	
yellow	4	10^4	
green	5	10^5	
blue	6	10^6	
violet	7	10^7	
gray	8	10^8	
white	9	10^9	
gold		10^{-1}	$\pm 5\%$
silver		10^{-2}	$\pm 10\%$

sample problem

Given a resistor having the color bands shown in the diagram, determine the value and tolerance of the resistor.

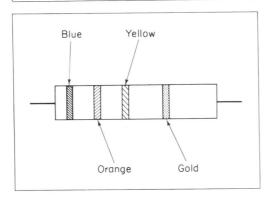

solution

Band A = 6
Band B = 3
Band C = 10⁴
Band D = ± 5% tolerance
The value is 63×10^4 ohms or 630 kΩ ± 5%

sample problem

A 50-ohm resistor is required to repair a radio. What color code is required if the tolerance is 10 percent?

solution

The first digit is five; therefore, the color is green. The second digit is zero; therefore, the color is black. The multiplier must be zero; therefore, the color is black. The color code required, therefore, is A = green, B = black, C = black, and D = silver.

semiconductor resistors

The semiconductor elements used in the manufacture of semiconductor devices are usually silicon and germanium. Consider the silicon atom shown

in Fig. 2–4. The nucleus contains 14 protons, each with an electrical charge
of +1 and a number of orbital shells containing a total of 14 electrons with
an electrical charge of −1.

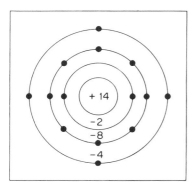

figure 2–4
silicon atom

The four outer electrons are loosely tied to the nucleus and can readily
be removed from their orbits by some adequate external force. If an atom
loses one of these electrons, then the number of protons exceeds the number
of electrons, and the atom has an electrical charge of +1. An atom that
has lost an electron is defined as an ionized atom and is called a *positive ion*.
An atom that has gained an electron in its outer shell is also an ionized atom
and is called a *negative ion*, since the number of electrons exceeds the number
of protons.

The structure of the silicon atom is a crystal lattice type, as shown
in Fig. 2–5.

The formation of this lattice structure permits two atoms to share
each of the four outer shell electrons associated with each atom. This sharing
of electrons is called a *covalent bond*. Although the four outer electrons of
the silicon atom can be readily dislodged by an external force, once they are
bound in the lattice structure, they become extremely difficult to remove.
Consequently, intrinsic or pure single-crystal silicon behaves as an insulator.

electrons and holes

The application of thermal energy to the lattice structure causes some
bonds between adjacent silicon atoms to be ruptured. This allows free electrons
to wander in random motion through the lattice. When a bond is broken,
a hole is left, due to one electron's leaving its orbit. The hole, therefore, has
a net positive charge of +1.

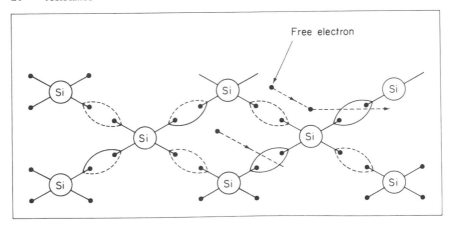

figure 2–5
crystalline structure with free electron in random motion

As temperature is increased, the number of broken bonds increases respectively, creating a larger number of electron hole pairs scattered throughout the lattice structure. Once an electron has escaped from its orbit, it wanders randomly through the crystal structure. As the electron approaches the vicinity of a hole, the positive charge attracts the electron, causing the electron to leap into the hole. As a result, both the electron and the hole disappear as free charges within the lattice structure.

A hole or positive charge also moves through the lattice in a somewhat analogous manner to the electron. The formation of a hole affects electrons in the adjacent covalent bond. In a vibrating crystal, the bonds are relatively loose, and therefore it is possible for the hole to *rob* an electron from the bond. Note that the hole has not neutralized but has merely moved to an adjacent atom. In this manner, the hole moves randomly throughout the crystal.

p-type silicon

The injection of boron atoms into the intrinsic silicon forms a *p*-type silicon. The element boron contains three electrons in its outer shell, so it can form covalent bonds with three adjacent silicon atoms, as shown in Fig. 2–6.

Since the fourth electron of boron is missing, a hole exists in the lattice structure. This hole attracts an electron from the adjacent atoms moving randomly throughout the material. Regardless of the number and type of impurities injected, the material itself is electrically neutral.

If the number of impurity atoms injected into the crystal exceeds the number of thermally created holes, the hole flow existent within the silicon material is essentially due to the impurity contamination. The resistivity of the material is less sensitive to temperature variation, since hole flow is controlled by the impurity atoms. Consequently, contaminated silicon can be used as a *resistor*, providing the impurity concentration or doping is adequate.

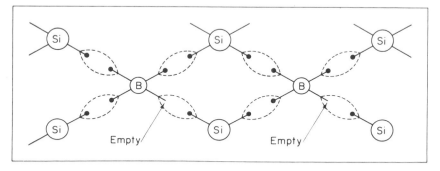

figure 2–6
boron atoms injected into silicon to form *p*-type silicon

Impurity atoms having three electrons in the outer shell are called *acceptor* atoms, because such electrons will accept electrons from the silicon crystal structure. Silicon atoms contaminated with acceptor atoms are called "*p*-type silicon." The application of a battery potential to the *p*-type silicon will cause holes to flow toward the negative terminal and free electrons to flow towards the positive terminal. Since there are more holes than free electrons, the hole flow is called the *majority carrier conduction*, and the electron flow is called the *minority carrier flow*.

n-type silicon

The injection of an impurity such as phosphorus containing five electrons in its outer shell yields "*n*-type silicon." In this case, only four of the electrons can form covalent bonds with the silicon atoms. The fifth electron is free to move randomly throughout the crystal structure as shown in Fig. 2–7.

Under these conditions, the crystal structure has an excess of electrons and is called "*n*-type silicon." Note that the element remains electrically neutral, since the free electron motion leaves a positive phosphorus ion behind. Impurity atoms that produce *n*-type silicon are called donor atoms because they donate free electrons to the material. The *n*-type material acts in a manner analogous to the *p*-type silicon; that is, the majority carriers are electrons and the minority carriers are holes.

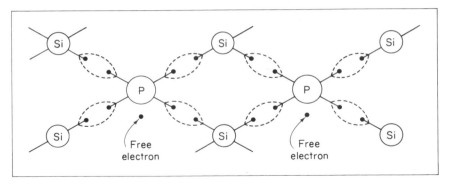

figure 2–7
phosphorus atoms injected into silicon to form *n*-type silicon

diffused silicon resistors

The resistance value of silicon is directly proportional to the resistivity and length of the material and inversely proportional to the area. The larger the impurity concentration, the lower the resistivity, with a consequent decrease in the resistance per unit length of the material.

For any given number of impurities within the material, the resistance value becomes dependent on the geometry of the structure.

An integrated circuit resistor is manufactured by diffusing an extremely narrow strip of doped material into the substrate of oppositely doped silicon, as shown in Fig. 2–8.

Connections are made to opposite ends of the diffused stripe, so that only the resistance of the stripe material becomes the measured circuit value.

The term *sheet resistance* is an often used expression in integrated circuit technology. Sheet resistance is defined as the resistance of a square of material of a given uniform thickness and is measured in ohms per square. This unit

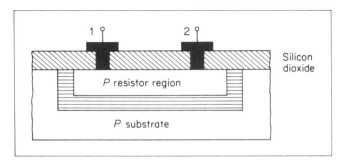

figure 2–8
a monolithic resistor

of measurement is based on the relationship that regardless of the size of the square, $l = w$, and the resistance of the material is then equal to the resistivity per unit thickness. Thus,

$$R = \rho \frac{l}{A} \text{ ohms}$$

and

$$R = \rho \frac{l}{t \times w} \text{ ohms}$$

$$R = R_s \frac{l}{w}$$

Since $l = w$, then $R = R_s$ (sheet resistance).

Intrinsic or pure silicon acts as an insulator and is valueless for resistor usage. In order properly to utilize the semiconductor material, impurities must be injected to introduce additional charge carriers. The temperature sensitivity of silicon can be reduced if the injected charge carriers outnumber the charge carriers provided within the material by thermal agitation.

Typical values for R_s range from 1 Ω/square to 250 Ω/square. For example, a base diffused resistor stripe 1 mil wide and 5 mils long contains five 1 mil $\times$ 1 mil squares with a sheet resistance of 100 Ω/square and has its ohmic value calculated by

$$R = 5 \times 100 = 500 \ \Omega$$

problems

1. Find the resistance of a copper wire no. 8 AWG that is 2500 ft long at 20°C.

2. What is the area of a copper wire in circular mils that is 500 ft long and has a resistance of 10 ohms?

3. What is the length of a copper wire that has a diameter of 0.0312 in. and a resistance of 4 ohms?

4. Find the resistance of a copper wire 50 yd in length and 0.0045 in. in diameter. Assume that the temperature is 20°C.

5. What is the resistance of 500 ft of no. 10 AWG copper wire at 20°C?

6. The resistance of a copper wire is 2 ohms at 10°C. What is the resistance of the wire at 70°C?

7. What length of copper wire is required for a resistance of 200 ohms at 150°C? The type of wire is no. 22 AWG.

8. The resistance of a copper wire is 10 ohms at 40°C. What is the resistance of the wire at −40°C?

9. If the resistance of a silver wire is 5 ohms at −30°C, what is its resistance at −5°C?

10. Find the resistance of a copper wire at 70°C if its resistance at 20°C is 30 ohms.

11. A copper wire measures 45 ohms at 20°C. If the hot resistance is 60 ohms, what is the rise in temperature?

12. What is the resistance of 100 ft of 22 gauge nickel wire at 200°C?

13. What are the ohmic values and the tolerance of the resistors given by the color band?

	A	B	C	D
(a)	brown	green	blue	gold
(b)	green	yellow	orange	gold
(c)	blue	red	yellow	silver
(d)	red	green	red	gold
(e)	yellow	gray	orange	silver
(f)	gray	white	gold	silver
(g)	orange	violet	yellow	gold
(h)	violet	green	orange	silver

14. Specify the color bands for each of the following resistors.

			A	B	C	D
(a)	100 Ω	± 10%				
(b)	910 kΩ	± 10%				
(c)	47 kΩ	± 10%				
(d)	560 kΩ	± 10%				
(e)	1 MΩ	± 10%				
(f)	0.25 Ω	± 10%				
(g)	2.6 Ω	± 10%				
(h)	4.7 MΩ	± 10%				

3

current, voltage, and power

circuit symbols

In the analysis of electrical circuits, the various electrical quantities are represented by symbols. For example, Fig. 3–1 illustrates a few of the more common symbols.

A battery having a voltage E and a polarity as shown in Fig. 3–1(a) demonstrates an emf source of energy. The symbol for a resistor can be either as shown in Fig. 3–1(b) or (c), depending on whether the resistor is

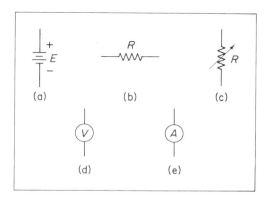

figure 3–1
electrical symbols: (a) battery, (b) resistor, (c) variable resistor,
(d) voltmeter, (e) ammeter

fixed or variable. The measurement of voltage is performed by a voltmeter (d). If the measurement of current is required, the ammeter (e) performs that function.

It should be noted that an electrical circuit diagram is a symbolic diagram and should not be considered a physical layout of the system. The length of wires shown in the symbolic diagrams bears no relationship to the actual length in the system. The circuit diagram is constructed for ease and simplicity of analysis.

Ohm's law

Sometime in the early part of the nineteenth century Ohm[1] noted that a relationship between voltage and current existed in a resistor. Consider the circuit shown in Fig. 3–2.

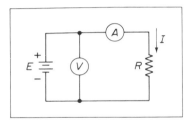

figure 3–2
circuit used for Ohm's law

The application of an emf E permits a current I to flow through the resistor. Doubling the voltage causes the current to double. Ohm's law states that the resistance of a circuit element is directly proportional to the voltage existing across the element and inversely proportional to the current flowing through the element. Consequently, the mathematical relationship called Ohm's law is stated as

$$R = \frac{E}{I} \text{ ohms} \qquad (3–1)$$

where E is in volts, I is in amperes, and R is in ohms. Several illustrative problems will demonstrate the theory.

[1]George Simon Ohm, a German physicist, developed Ohm's law in 1827.

problem 1

A current of 5 amperes flows through a 10-ohm resistor. Determine the potential difference across the resistor.

solution

$$E = IR$$
$$E = 5 \times 10 = 50 \text{ volts}$$

problem 2

What is the current flowing through a 30-ohm resistor that has a potential difference of 300 volts across it?

solution

$$I = \frac{E}{R}$$

$$I = \frac{300}{30} = 10 \text{ amperes}$$

problem 3

What is the ohmic resistance of a toaster that draws 6 amperes from a 120-volt line?

solution

$$R = \frac{E}{I} = \frac{120}{6} = 20 \ \Omega$$

power

Power is the rate of doing work. Mathematically, power is defined as joules per second, or

$$\text{power} = \frac{\text{work}}{\text{time}} \frac{\text{joules}}{\text{second}}$$

In electrical circuit theory, voltage or potential difference is equal to the work applied in moving a test charge. Current is defined as the flow of charge carriers per unit time. Then

$$EI = \frac{\text{work}}{\text{charge}} \frac{\text{charge}}{\text{time}} = \text{power (watts)}$$

Power in an electrical circuit is defined in terms of a unit called the *watt*. Thus, when the current flowing through the circuit is 1 ampere and the voltage across two terminals within the circuit is 1 volt, the power dissipated across the two terminals is 1 joule per second, or 1 watt.

Substituting Ohm's law into the power relationship yields the following.

$$P = EI = E\frac{E}{R} = I(IR) \text{ watts} \qquad (3\text{--}2)$$

An illustration of the relationship between Ohm's law and the power relationships is shown diagrammatically in Fig. 3–3.

It should be noted that power developed in a resistor converts electrical

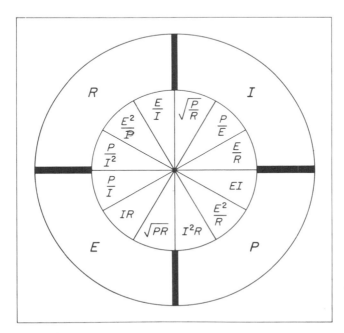

figure 3–3
power, voltage and current interrelationships

energy into heat. The greater the power applied, the hotter the resistor becomes. Each resistor has a heat dissipation or wattage limitation that should not be exceeded for safety and proper operation.

efficiency

In any electrical circuit, the power applied must approximately equal the power dissipated in the circuit. Thus,

$$P_i = P_0 + P_{diss}$$

where P_i is the total input or applied power, P_0 is the output power dissipated across the resistor, and P_{diss} is the power dissipated in the wires. The efficiency of a circuit is defined as the ratio of the output power to the applied input power. This relationship is expressed by

$$\% \ \eta \ (efficiency) = \frac{P_0}{P_i} \ 100\% \tag{3-3}$$

Efficiencies can be applied to any system. For example, input electrical power may energize a motor to produce a mechanical output power. The conversion of electrical power to mechanical power requires the following mathematical relationship:

$$1 \text{ horsepower (1 hp)} = 746 \text{ watts}$$

and

$$550 \ \frac{\text{ft-lb}}{\text{sec}} = 1 \text{ hp}$$

It should be noted that 1 ft-lb is the amount of work performed in lifting a 1-lb weight through a distance of 1 ft.

---

sample problem 1

An electric motor operating from a 220-v source requires a current of 10 amperes. The motor operates with an efficiency of 80 percent. What is the output of the motor in horsepower?

---

solution

step 1: Calculate the input power.

$$P_i = EI = 220 \times 10 = 2.2 \text{ kw}$$

step 2:   Calculate the output power.

$$P_0 = \eta P_i = 0.8 \times 2.2 \text{ kw}$$
$$P_0 = 1.76 \text{ kw}$$

step 3:   Determine the horsepower output.

$$Hp = \frac{1.76 \text{ kw}}{0.746} = 2.35 \text{ hp}$$

---

sample problem 2

A 6-hp motor operates at an efficiency of 75 percent. What is the current input to the motor if the applied voltage is 220 v?

---

solution

step 1:   Calculate P_i.

$$\eta = \frac{P_0}{P_i} 100\%$$

$$P_i = \frac{6}{0.75} = 8 \text{ hp}$$

step 2:   Calculate P_i in watts.

$$P_i = 8 \times 746 = 5.968 \text{ kw}$$

step 3:   Calculate I.

$$I = \frac{P_i}{E} = \frac{5.968}{220}$$
$$I = 27.2 \text{ amperes}$$

energy

Power is defined as the rate of performing work. If a certain amount of power is developed over a specified length of time, then energy is expended. Consequently, energy can be defined as the ability or capacity to do work. Thus, the amount of energy expended in accomplishing useful work is equal to the power multiplied by time. A common unit of energy is the joule. Thus, 1 joule is equal to 1 watt-second. Note that the watt-second is then defined as the unit of electrical energy. Since electrical energy can be purchased, the watt-second unit was found to be too small for practical considerations. The generally accepted unit in commercial applications is the kilowatt hour. The various mathematical conversion formulas are

$$1 \text{ kwh} = 3.6 \times 10^6 \text{ joules}$$
$$1 \text{ kwh} = 1.34 \text{ hp-hr}$$
$$1 \text{ kwh} = 2.653 \times 10^6 \text{ ft-lb-hr}$$

---

sample problem

The cost of electrical energy is based on the consumption of power per hour, or kilowatt hours. Consider a 100-watt lamp that is operating continuously for 1 year. If the cost is 3.2 cents per kilowatt hour, determine the cost of the lamp operation over 1 year.

---

solution

step 1: Calculate the number of kilowatt hours per year.

$$N \text{ kwh} = \frac{100}{1000} \times 365 \times 24$$

$$N \text{ kwh} = 876 \text{ kwh}$$

step 2: Determine the cost in dollars.

$$\frac{876}{100} \times 3.2 = \$28.03$$

problems

1. The voltage across an electric toaster is 240 volts, and the current taken by the toaster is 8 amperes. Calculate the resistance of the toaster.

2. The resistance of a relay coil is 10,000 ohms. The relay operates with a hold in current of 50 ma. At what voltage does the relay pull in?

3. A dc relay coil has 12 volts applied and a resistance of 500 ohms. What current will the coil take?

4. The resistance of an incandescent lamp is 1000 ohms. When a voltage of 120 volts is applied, determine the current.

5. Determine the potential difference across a 15-ohm resistor that has a current of 10 ma flowing through it.

6. What is the voltage drop across a 1-megohm resistor if the current flow through the resistor is 15 μa?

7. A heating element has a resistance of 200 ohms. Find the current through the element if 110 volts is applied across it.

8. An electric clock has an internal resistance of 7 kilohms. Find the current through the clock if a voltage of 210 volts is applied across it.

9. A voltmeter has an internal resistance of 20,000 ohms. A current of 6 ma is flowing through the meter. What is the voltage reading?

10. A 50-ohm resistor is rated at 5 watts. What is the permissible current that can flow through it? What is the maximum possible voltage that can be placed across it?

11. A 700-watt heater is rated at 120 volts. What is the current rating of the heater?

12. The current in a 5600-ohm resistor is 25 ma. What is its voltage and power rating?

13. A 4-kΩ resistor dissipates 2000 joules in 10 minutes. Determine its resistance and current rating.

14. A 50-watt soldering iron is rated at 120 volts. Determine the resistance and current rating of the iron.

15. If 5500 joules of energy is absorbed by a 5-MΩ resistor in 11 minutes, determine the current rating and the voltage that can exist across the resistor.

16. A resistor of 20 ohms has a charge of 500 coulombs per second flowing through it. How much power is dissipated in the resistor?

17. What is the voltage rating and resistance of an electric washer that has a 1000-watt rating and draws a current of 7.5 amperes?

18. What is the output of a 5-hp motor in foot-pounds per hour?

19. Determine the efficiency of a system that has an output of 0.5 hp and an input power of 400 watts.

20. If an electric motor has an efficiency of 85 percent and, operating from a 220-volt line, delivers 4.5 hp, what input current does the motor draw?

4

dc circuits

series circuits

In the foregoing chapters, the concepts of current, voltage, and power were considered as numerical quantities bearing no relationship to circuits. The properties of dc circuits form the basic foundation for the understanding of electrical and electronic circuit applications.

Consider the definition of Ohm's law. The circuit used to describe this law is a series circuit. An *electric circuit* may consist of any number of circuit elements connected in series, parallel, or any combination whatsoever, provided that at least one closed path exists wherein current flows. A *series circuit* is defined as a circuit in which the current flow is the same constant measured anywhere in the circuit. Figure 4–1 illustrates a simple series circuit containing a battery and three resistors.

The potential difference across each resistor can be determined by using Ohm's law. Thus,

$$E_1 = IR_1$$
$$E_2 = IR_2$$
$$E_3 = IR_3$$

The sum total of the potential differences must be equal to the applied potential. Thus,

$$E = E_1 + E_2 + E_3$$

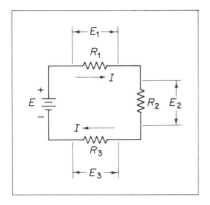

figure 4–1
simple series circuit

Substituting, we obtain

$$E = IR_1 + IR_2 + IR_3$$

or

$$E = I(R_1 + R_2 + R_3)$$

The total or equivalent resistance in a series circuit is equal to the sum of the resistances connected in series. Thus,

$$R_T = R_1 + R_2 + R_3$$

Applying this conclusion to the simple series circuit shown in Fig. 4–2 yields the following results:

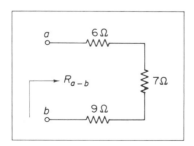

figure 4–2
series circuit

The total resistance of the circuit is $R_{a-b} = R_T$. Thus,

$$R_{a-b} = 6 \text{ ohms} + 7 \text{ ohms} + 9 \text{ ohms} = 22 \text{ ohms}$$

It should be noted that when the total resistance of the circuit R_T is multiplied by I^2, the resultant power equation is

$$I^2 R_T = I^2 R_1 + I^2 R_2 + I^2 R_3$$

Assume that the current flowing in the above circuit is 1 ampere. The power dissipated across each resistor is

$$P_1 = 1^2 \times 6 = 6 \text{ watts}$$

$$P_2 = 1^2 \times 7 = 7 \text{ watts}$$

$$P_3 = 1^2 \times 9 = 9 \text{ watts}$$

The total power is equal to the sum of the power drops in the circuit. Thus,

$$P_T = P_1 + P_2 + P_3$$

$$P_T = 22 \text{ watts.}$$

---

sample problem

The following circuit is given. Determine the line current and the power dissipation across each resistor.

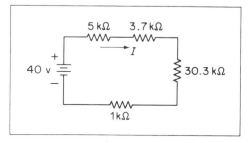

solution

step 1:   Calculate R_T.

$$R_T = 5000 \text{ ohms} + 3700 \text{ ohms} + 30{,}300 \text{ ohms} + 1000 \text{ ohms}$$
$$R_T = 40{,}000 \text{ ohms}$$

step 2:   Calculate I.

$$I = \frac{E}{R_T} = \frac{40}{40{,}000} = 1 \text{ ma}$$

step 3:   Calculate P_T.

$$P_1 = I^2 R_1 = (10^{-3})^2 \, 5 \times 10^3 = 5 \text{ mw}$$
$$P_2 = I^2 R_2 = (10^{-3})^2 \, 3.7 \times 10^3 = 3.7 \text{ mw}$$
$$P_3 = I^2 R_3 = (10^{-3})^2 \, 30.3 \times 10^3 = 30.3 \text{ mw}$$
$$P_4 = I^2 R_4 = (10^{-3})^2 \, 10^3 = 1 \text{ mw}$$
$$P_T = 40 \text{ mw}$$

---

sample problem

In the given circuit, determine the current I and the power across each resistor.

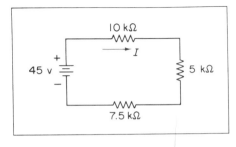

solution

step 1:   Calculate I.

$$I = \frac{E}{R_T} = \frac{45}{22.5 \times 10^3}$$

$$I = 2 \text{ ma}$$

step 2:   Calculate the individual powers.

$$P_1 = I^2 R_1 = (2 \times 10^{-3})^2 \, 10^4$$
$$P_1 = 40 \text{ mw}$$
$$P_2 = I^2 R_2 = (2 \times 10^{-3})^2 \, 5 \times 10^3$$
$$P_2 = 20 \text{ mw}$$
$$P_3 = I^2 R_3 = (2 \times 10^{-3})^2 \, 7.5 \times 10^3$$
$$P_3 = 30 \text{ mw}$$

The total input power is equal to the sum of the individual powers. Thus,

$$P_i = P_1 + P_2 + P_3 = 20 + 30 + 40 = 90 \text{ mw}$$

Note also that the input voltage is split up among the resistors in the following manner: 20 volts across the 10-kilohm resistor, 10 volts across the 5-kilohm resistor, and 15 volts across the 7.5-kilohm resistor. Gustav R. Kirchhoff discovered this fundamental relationship, which today is known as Kirchhoff's voltage law. This law can be stated as follows: The algebraic sum of all voltages taken in a closed loop in sequential order must equal zero. Symbolically, we have

$$\Sigma_\sigma\, E = 0$$

Another way of stating this same law is as follows: The applied voltage must equal the sum of all the potential drops in any complete electric circuit.

It is necessary at this point to define the polarity of voltage across a resistor due to current flow. The conventional method assumes that current flows out of a positive battery terminal entering resistances at a positive terminal and flowing out of the negative terminal of the resistances. The arrowhead normally denotes the direction of current flow. This convention will be used throughout this text.

Refer to Fig. 4–3. Note that the circuit terminals are marked a, b, and c, respectively. The application of Kirchhoff's voltage law proceeds in the following manner. Start with terminal a and look towards terminal b. An

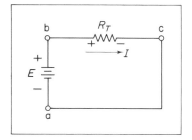

figure 4–3
basic series circuit

emf of E volts exists and is considered a *voltage rise* from a to b. Consequently, a potential rise is considered a positive value, whereas a potential drop (from positive to negative) is a negative value. Thus, in closed loop $abca$, the following equation can be written.

$$+E - IR_T = 0$$

Another example is shown in Fig. 4–4. The Kirchhoff's law equation in a closed loop $abcdea$ is

$$+E - IR_1 - IR_2 - IR_3 - IR_4 = 0$$

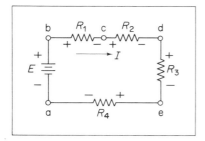

figure 4–4
series circuit

If the battery polarity is reversed and the current direction is assumed to be as shown in Fig. 4–5, the resultant closed-loop equation is

$$-E - IR_1 - IR_2 - IR_3 - IR_4 = 0$$

Note that the resultant value of current will be negative. The negative sign merely indicates that the current direction assumed is incorrect and should be reversed.

voltage dividers

In many practical applications, a series resistor may be inserted between source and load to reduce the input voltage to some desired value. The problem of applying a 24-volt source to a 10-volt, 0.2-ampere bell illustrates the theory. Note that a resistor must be placed in series between the source and load, as shown in Fig. 4–6.

Since 14 volts must be dropped across R and the line current is 0.2 ampere, the value of R is

$$R = \frac{14}{0.2} = 70 \, \Omega$$

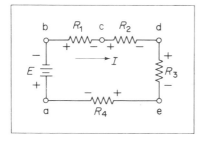

figure 4–5
series circuit

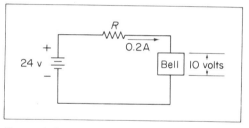

figure 4–6
series resistor between source and load

The power dissipated across R is

$$P = I^2R = (0.2)^2\ 70 = 2.8 \text{ watts}$$

In any series circuit containing two or more resistors, the voltage applied may be split up across the individual resistors. Consequently, the resistors being used to reduce voltage are known as *voltage dividers*. It is evident, therefore, that the current in a series circuit is the same throughout and that the resistor having the largest ohmic values will develop the largest voltage across its terminals.

The general mathematical development refers to Fig. 4–7. Note that the voltage E_2 is

$$E_2 = IR_2$$

and

$$I = \frac{E}{R_1 + R_2}$$

Substituting yields the following equation:

$$E_2 = E\frac{R_2}{R_1 + R_2}$$

(4–1)

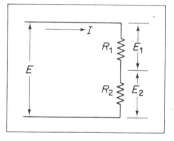

figure 4–7
voltage divider circuit

If the voltage across R_1 is desired, the procedure is the same. Thus,

$$E_1 = E \frac{R_1}{R_1 + R_2} \qquad (4\text{-}2)$$

Consequently, the voltage divider rule can be expressed as follows: Consider the resistor R_x across which the voltage E_x is desired. Then

$$E_x = E \frac{R_x}{R_T} \qquad (4\text{-}3)$$

where

R_T = sum of all the resistors in series.
R_x = resistance of the resistor across which the voltage is desired.
E_x = voltage across the resistor of interest.
E = applied voltage.

An example will demonstrate the theory.

---

sample problem

Given the circuit shown, find E_1, E_2, and E_3
where $E = 42$ volts $R_2 = 2$ kΩ
        $R_1 = 4$ kΩ $R_3 = 1$ kΩ

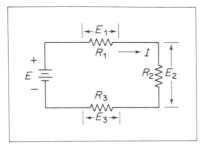

solution

step 1:   Calculate E_1.

$$E_1 = E \frac{R_1}{R_T}$$

$$E_1 = 42 \tfrac{4}{7} = 24 \text{ volts}$$

step 2:   Calculate E_2.

$$E_2 = E \frac{R_2}{R_T}$$

$$E_2 = 42 \tfrac{2}{7} = 12 \text{ volts}$$

step 3:   Calculate E_3.

$$E_3 = E \frac{R_3}{R_T}$$

$$E_3 = 42 \tfrac{1}{7} = 6 \text{ volts}$$

A check on the solution is to sum up the voltages, which must equal the applied voltage of 42 volts. Thus,

$$\Sigma_\circlearrowleft E = 0$$

$$0 = +E - E_1 - E_2 - E_3$$

or

$$42 \text{ v} = (12 + 24 + 6) \text{ volts}$$

$$42 \text{ v} = 42 \text{ v} \quad \text{(The results check.)}$$

Voltage sources may be connected in series for increasing or decreasing output potentials. Refer to Fig. 4–8. It is evident that the same current flows out of both sources.

Note that in Fig. 4–8(a) the output voltage E_{0_1} is greater in magnitude than either E_1 or E_2. Thus,

$$E_{0_1} = E_1 + E_2$$

Consequently, the sources are series aiding, since both batteries maintain the flow of current to a load. In Fig. 4–8(b) the output voltage is less than

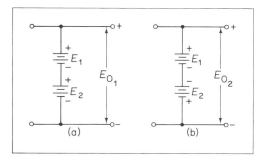

(a)   (b)

figure 4–8
series voltage sources: (a) increasing, (b) decreasing

either source by itself. If the positive terminal of E_1 is considered as the reference terminal, then

$$E_{0_2} = E_1 + (-E_2)$$

In this case the sources are series opposing or bucking each other, since the net output voltage E_{0_2} is less than either E_1 or E_2.

parallel circuits

A circuit is defined as a parallel circuit when two or more circuit elements are so connected that the same voltage exists across each element. A number of resistors are connected as shown in Fig. 4–9.

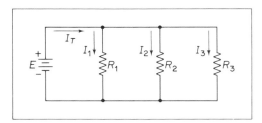

figure 4–9
parallel circuit

The applied voltage being the same across each resistor, the individual branch currents are, by Ohm's law,

$$I_1 = \frac{E}{R_1} \; ; \qquad I_2 = \frac{E}{R_2} \; ; \qquad I_3 = \frac{E}{R_3}$$

Kirchhoff, investigating the phenomena of parallel circuitry, noted that the sum total of all the branch currents must be equal to the source current. He formulated the current law, which says in essence that the algebraic sum of all currents entering or leaving a node[1] must be equal to zero.

$$\Sigma I = 0$$

The application of Kirchhoff's current law to the parallel circuit yields

$$I_T = I_1 + I_2 + I_3$$

[1]A node is a terminal in a circuit where current enters or leaves.

Substituting yields

$$I_T = \frac{E}{R_1} + \frac{E}{R_2} + \frac{E}{R_3} = E \left(\frac{1}{R_1} + \frac{1}{R_2} + \frac{1}{R_3} \right)$$

Since Ohm's law states that $I_T = \dfrac{E}{R_T}$, then

$$\frac{1}{R_T} = \frac{1}{R_1} + \frac{1}{R_2} + \frac{1}{R_3}$$

A general equation for the evaluation of the total resistance of a parallel circuit is

$$\boxed{\frac{1}{R_T} = \frac{1}{R_1} + \frac{1}{R_2} + \frac{1}{R_3} + \cdots + \frac{1}{R_n}}$$ (4–4)

An illustrative problem will demonstrate the theory.

sample problem

Given the circuit shown, determine the value of R_T and I_T
where $E_1 = 15$ v $R_2 = 8$ kΩ
$E_2 = 45$ v $R_3 = 6$ kΩ
$R_1 = 12$ kΩ

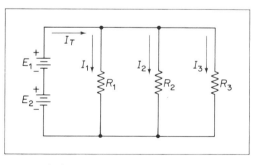

solution

step 1:   Calculate R_T.

$$\frac{1}{R_T} = \frac{1}{R_1} + \frac{1}{R_2} + \frac{1}{R_3}$$

$$\frac{1}{R_T} = \left(\frac{1}{12} + \frac{1}{8} + \frac{1}{6}\right)10^{-3}$$

$$\frac{1}{R_T} = \left(\frac{2+3+4}{24}\right)10^{-3}$$

$$R_T = 2.67 \text{ k}\Omega$$

step 2:   Calculate I_T.

$$I_T = \frac{E}{R_T}$$

$$I_T = \frac{60}{2.67 \times 10^3}$$

$$I_T = 22.5 \text{ ma}$$

Using Kirchhoff's current law as a check, solve for I_1, I_2, and I_3, respectively. Thus,

$$I_1 = \frac{60}{12 \times 10^3} = 5 \text{ ma}$$

$$I_2 = \frac{60}{8 \times 10^3} = 7.5 \text{ ma}$$

$$I_3 = \frac{60}{6 \times 10^3} = 10 \text{ ma}$$

$$I_T = 22.5 \text{ ma} \quad \text{(The results check.)}$$

Note that since the individual branch currents can be determined, the power dissipated across each resistor can be calculated. Thus,

$$P_1 = I_1^2 R_1 = EI_1$$
$$P_2 = I_2^2 R_2 = EI_2$$
$$P_3 = I_3^2 R_3 = EI_3, \text{ etc.}$$

The input supply power must equal the sum of the power dissipated in the circuit elements. Thus,

$$P_T = I_1^2 R_1 + I_2^2 R_2 + I_3^2 R_3 + \cdots + I_n^2 R_n$$

or

$$P_T = P_1 + P_2 + P_3 + \cdots + P_n$$

Continuing with the sample problem and calculating the various powers yields

$$P_1 = I_1^2 R_1 = (5 \times 10^{-3})^2 (12 \times 10^3) = 300 \text{ mw}$$

$$P_2 = I_2^2 R_2 = (7.5 \times 10^{-3})^2 (8 \times 10^3) = 450 \text{ mw}$$

$$P_3 = I_3^2 R_3 = (10 \times 10^{-3})^2 (6 \times 10^3) = 600 \text{ mw}$$

$$P_T = I_T^2 R_T = (22.5 \times 10^{-3})^2 (2.67 \times 10^3) = 1350 \text{ mw}$$

To check these results, we have

$$P_T = P_1 + P_2 + P_3 = (300 + 450 + 600) \text{ mw}$$

$$1350 \text{ mw} = 1350 \text{ mw} \quad \text{(The results check.)}$$

current dividers

The conductance G of a resistor is defined as the reciprocal of resistance. Consequently, since

$$\frac{1}{R_T} = \frac{1}{R_1} + \frac{1}{R_2} + \frac{1}{R_3} + \cdots + \frac{1}{R_n}$$

then

$$G_T = G_1 + G_2 + G_3 + \cdots + G_n$$

The total conductance G_T in a parallel circuit is larger than any one of the individual branch conductances. Consequently, R_T, being the reciprocal of G_T, will be smaller than any one of the individual branch resistances. In a series circuit the total resistance is greater than any one of the individual resistances.

A common problem is to find the total resistances of just two resistors in parallel. Thus,

$$\frac{1}{R_T} = \frac{1}{R_1} + \frac{1}{R_2}$$

Then

$$R_T = \frac{R_1 R_2}{R_1 + R_2}$$

This equation states that the total resistance of two resistors in parallel is the product divided by the sum of the two resistors. In a parallel circuit, the following relationships are valid.

$$E = I_1 R_1 = I_2 R_2 = I_T R_T$$

It is often desirable to be able to determine the branch current in terms of the source current.

$$I_1 R_1 = I_T \left(\frac{R_1 R_2}{R_1 + R_2} \right)$$

and

$$I_1 = I_T \left(\frac{R_2}{R_1 + R_2} \right)$$

$$I_2 = I_T \left(\frac{R_1}{R_1 + R_2} \right)$$

(4–5)

Examining these two equations carefully indicates the general form of the current divider relationship. In a parallel circuit of two resistors, *the current in the resistor of interest is equal to the source or input current multiplied by the opposite resistor divided by the sum of the two resistors.*

sample problem

Given the circuit shown, determine I_1 and I_2
where $R_1 = 40\ \Omega$
$\qquad R_2 = 60\ \Omega$

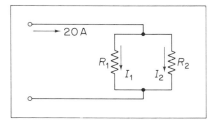

solution

step 1: Calculate I_1.

$$I_1 = I_T \left(\frac{R_2}{R_1 + R_2} \right)$$

$$I_1 = 20 \left(\frac{60}{40 + 60} \right)$$

$$I_1 = 12 \text{ amperes}$$

step 2:   Calculate I_2.

$$I_2 = I_T \left(\frac{R_1}{R_1 + R_2} \right)$$

$$I_2 = 20 \left(\frac{40}{60 + 40} \right)$$

$$I_2 = 8 \text{ amperes}$$

series parallel circuits

Most practical circuits consist of complex networks containing series parallel connected elements. The analysis of such circuits is an extension of the methods and principles underlying series and parallel circuits.

The circuit shown in Fig. 4–10 is a series parallel circuit. Note that R_1 is in series with the parallel combination of R_2 and R_3. The current through R_1 is the total or I_T, which must equal the sum of the currents flowing through R_2 and R_3. The voltage E is equal to E_1 and E_2.

An illustrative problem will demonstrate the theory.

sample problem

The following circuit is given. Find: (a) I_1, I_2, and I_T, (b) E_1 and E_2, and (c) the power dissipated across each resistor
where $R =$ 6 kilohms
$\quad\quad\quad R_1 =$ 8 kilohms
$\quad\quad\quad R_2 =$ 12 kilohms

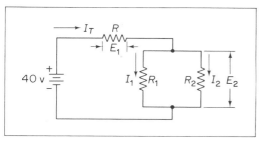

solution

step 1:   Calculate R_T.

$$R_T = R + \frac{R_1 R_2}{R_1 + R_2}$$

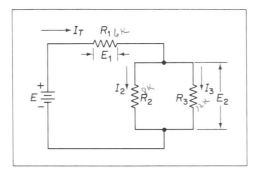

figure 4–10
series parallel circuit

$$R_T = \left(6 + \frac{8 \times 12}{8 + 12}\right)10^3$$

$$R_T = 10.8 \text{ kilohms}$$

step 2: Calculate the currents.

$$I_T = \frac{E}{R_T} = \frac{40}{10.8 \times 10^3}$$

$$I_T = 3.7 \text{ ma}$$

$$I_1 = I_T\left(\frac{R_2}{R_1 + R_2}\right) = 3.7\left(\frac{12}{8 + 12}\right)$$

$$I_1 = 2.22 \text{ ma}$$

$$I_2 = I_T\left(\frac{R_1}{R_1 + R_2}\right) = 3.7\left(\frac{8}{8 + 12}\right)$$

$$I_2 = 1.48 \text{ ma}$$

step 3: Calculate E_1 and E_2.

$$E_1 = I_T R = 3.7 \times 6$$
$$E_1 = 22.2 \text{ volts}$$
$$E_2 = I_2 R_2 = 1.48 \times 12$$
$$E_2 = 17.76 \text{ volts}$$

A check on this work is made as follows.
$$E = E_1 + E_2 = 22.2 + 17.76$$
$$40 = 39.96 \text{ volts}$$

The results are fairly close, and the answer is considered correct.

step 4:   Calculate the powers dissipated across each resistor.

$$P = I_T^2 R = (3.7 \times 10^{-3})^2 \, 6 \times 10^3$$
$$P = 82.14 \text{ mw}$$
$$P_1 = I_1^2 R_1 = (2.22 \times 10^{-3})^2 \, 8 \times 10^3$$
$$P_1 = 39.43 \text{ mw}$$
$$P_2 = I_2^2 R_2 = (1.48 \times 10^{-3})^2 \, 12 \times 10^3$$
$$P_2 = 26.29 \text{ mw}$$
$$P_T = I_T^2 R_T = (3.7 \times 10^{-3})^2 \, 10.8 \times 10^3$$
$$P_T = 148 \text{ mw}$$

A check on these results yields
$$P_T = P_1 + P_2 + P$$
$$148 \text{ mw} = (82.14 + 39.43 + 26.29) \text{ mw}$$
$$148 \text{ mw} = 147.86 \text{ mw}$$

(The difference is due to the squaring of minute error in voltage potentials.)

---

sample problem

Determine R_n, I_T, I_1, I_2, and the powers across each resistor for the circuit shown.

---

solution

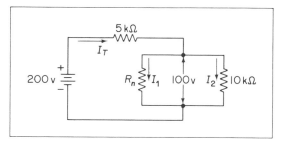

step 1:   Let $R_2 = 10$ kΩ, $E = 200$ v, $E_2 = 100$ v, and $R = 5$ kΩ. Then the voltage across R is equal to

$$E_1 = E - E_2 = 200 - 100 = 100 \text{ volts}$$

step 2:   Calculate I_T.

$$I_T = \frac{E_1}{R} = \frac{100}{5 \times 10^3}$$
$$I_T = 20 \text{ ma}$$

step 3:   Calculate I_2.

$$I_2 = \frac{E_2}{R_2} = \frac{100}{10 \times 10^3}$$

$$I_2 = 10 \text{ ma}$$

step 4:   Calculate I_1.

$$I_1 = I_T - I_2 = (20 - 10) \text{ ma}$$

$$I_1 = 10 \text{ ma}$$

step 5:   Calculate R_n.

$$R_n = \frac{E_2}{I_1} = \frac{100}{10 \times 10^{-3}}$$

$$R_n = 10 \text{ k}\Omega$$

step 6:   Calculate the power dissipation.

$$P = I_T^2 R = (20 \times 10^{-3})^2 \; 5 \times 10^3$$
$$P = 2 \text{ w}$$
$$P_1 = I_1^2 R_n = (10 \times 10^{-3})^2 \; 10 \times 10^3$$
$$P_1 = 1 \text{ w}$$
$$P_2 = 1 \text{ w}$$

A check on these results is
$$P_T = P + P_1 + P_2 = (2 + 1 + 1) = 4 \text{ w}$$
$$P_T = I_T^2 R_T = (20 \times 10^{-3})^2 \; 10 \times 10^3 = 4 \text{ w}$$

The results check, since 4 w = 4 w

problems

1. A 12.5-kilohm resistor, a 26.25-kilohm resistor, and a 1250-ohm resistor are connected in series across a 440-volt source. Determine the current flowing through each resistor, the potential difference across each resistor, and the power dissipation of each resistor.

2. Three resistors are connected in series across a 120-volt source. The second resistor is three times the value of the first resistor and the third resistor is eight times the value of the first resistor. The total current from the source is 0.2 ampere. What is the value of each resistor? What is the potential difference and the power dissipated across each resistor?

3. A resistor is placed in series with a supply to produce a voltage drop of 30 volts. The load resistor is rated at 660 watts when the voltage across the load is 110 volts. Determine the value of resistor R and the power dissipated across R.

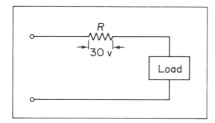

4. Three resistors are in series. The resistance of the first is 24 ohms, the power dissipation of the second is 50 watts, and the voltage drop across the third is 30 yolts. When the current is doubled, the voltage drop across the first resistor is 48 volts. Determine the value of the unknown resistors and determine the original supply voltage.

5. Three resistors are connected in series. The first is a 2-watt resistor of 10 kilohms. The second is a half-watt resistor of 2.5 kilohms. The third is a 5-watt resistor of 20 kilohms. Determine: (a) the maximum current the circuit can handle without exceeding rated power; (b) the input voltage, using the source current of part (a).

6. Four resistors are in series. They are, respectively,

10 kilohms— 3 watts	5 kilohms—2 watts
20 kilohms—10 watts	15 kilohms—6 watts

Determine: (a) the maximum source current that can be applied without exceeding rated power; (b) the source voltage, using the source current of part (a).

7. Determine the resistance and the power dissipation of a resistor that must be placed in series with a 100-ohm resistor having a 100-watt rating. The source voltage is 220 volts.

8. A series circuit of lamps consists of ten 6-watt lamps for use on 120 volts. Determine the current and resistance of each lamp.

9. An air-conditioning outlet is fused by 10 amperes and is rated at 750 watts. Determine the maximum possible resistance that can be placed across the outlet.

10. A lamp requires a current of 2.4 amperes at 120 volts for required output. A resistor is inserted in series with the lamp to drop the voltage from 120 volts to 90 volts. Determine the required value of resistor and its power rating.

11. Three resistors of 20 ohms, 60 ohms, and 30 ohms are connected in parallel. Determine the total resistance.

12. Three resistors of 80 kilohms, 60 kilohms, and 20 kilohms are connected in parallel. Determine the total resistance.

13. Four resistors of 65 kilohms, 130 kilohms, 260 kilohms, and 195 kilohms are connected in parallel. Determine the total resistance.

14. Four resistors of 75 ohms, 150 ohms, 300 ohms, and 600 ohms are connected across a 225-volt line. Determine the total resistance and the total current flow.

15. Two resistors are in ′parallel with a third resistor. The values of the known resistors are 200 kilohms and 150 kilohms. What is the value of third the resistor so that the total resistance is equal to 50 kilohms?

16. Four resistors are connected in parallel. The values for three of them are specified as
$$R_1 = 100 \text{ k}\Omega, \qquad R_2 = 100 \text{ k}\Omega, \qquad R_3 = 200 \text{ k}\Omega$$
Determine the resistance of the fourth resistor needed for the total resistance to be 33.3 kilohms.

17. Three resistors of 1000 ohms, 500 ohms, and 3000 ohms are connected in parallel. The total current is 10 ma. What is the applied voltage and what are the branch currents?

18. Two resistors of 30 ohms and 90 ohms are in parallel. The total current is 20 ma. What are the branch currents and the applied voltage?

19. Two resistors are in parallel, and the line current is 30 ma. The values of the resistors are 1 kilohm and 3 kilohms, respectively. What are the branch currents and the applied voltage?

20. The conductances of the branches of a parallel circuit are 125 microohms and 2 milliohms. If the line current is 50 ma, what are the branch currents and the voltage across the parallel combination?

21. Two resistors R_1 and R_2 are to be connected in parallel so that the current through R_1 is twice the value of current in R_2. The total resistance of the parallel combination is 240 ohms. Find R_1 and R_2.

22. Three lamps connected in parallel across a 120-volt source are rated at 60 watts, 75 watts, and 100 watts, respectively. What are the total input current, the individual branch currents, and the total resistance?

23. Three resistances are connected in parallel. A source current of 200 ma is applied to the network. The first resistance has a value of 4 kilohms. The second resistor has a current of 50 ma flowing through it. The third resistor has a voltage of 150 volts across it. Calculate the resistance of the second and third resistors.

24. What resistance must be placed in parallel with a 25-kilohm resistor for the total resistance to be 15 kilohms?

25. Using the current divider rule, find the unknown currents in the circuits shown in the figure.

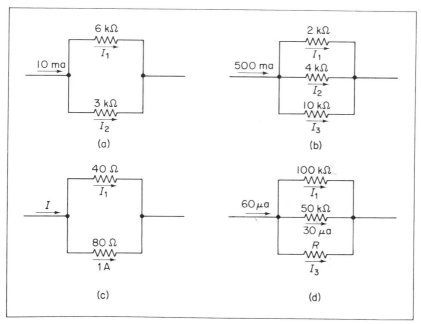

26. Solve for the unknown currents flowing in and the powers dissipated across each resistor in the circuits shown.

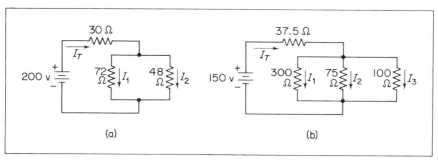

27. Solve for the unknown currents flowing in and the powers dissipated across each resistor in the circuits shown.

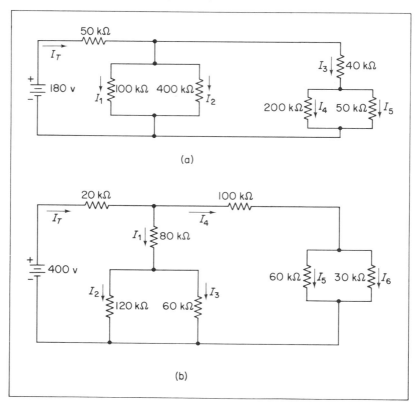

(a)

(b)

5

dc meters

introduction

In order to determine the performances of electrical and electronic equipment, it is often necessary to make some measurements. The equipment and the measuring techniques that are used for any measurement depend on the purpose of the measurement. The intent of this chapter is to present the principles and techniques used in the measurement of basic electrical devices. Most meters utilize one of the three basic types of meter movement, namely: (1) D'Arsonval, (2) electrodynamic, (3) iron vane.

D'Arsonval meter

The movement shown in Fig. 5–1 is used both for current and voltage measurements. The D'Arsonval movement consists of a coil with many turns of fine wire wound on an iron cylinder and mounted between the pole faces of a permanent magnet. The iron cylinder is free to rotate on pivots in either direction. A helical spring is used to hold the coil in a fixed position. The tension in the spring limits the turning motion of the coil.

The application of current to the coil results in the creation of a magnetic field around the coil. The magnetic fluxes generated by the coil and the permanent magnet interact. A torque is developed on the coil which causes the cylinder to rotate on the pivots. The cylinder rotates from the zero position until a counter torque is developed by the helical spring exactly equal to the magnetic torque. The pointer is then displaced from the

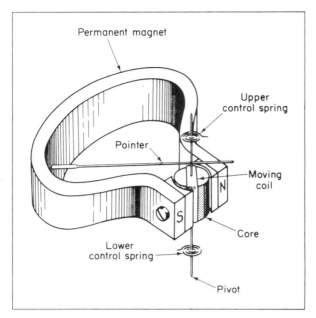

figure 5–1
arrangement of the basic parts of a D'Arsonval meter

origin by an amount directly proportional to the amount of current flowing through the coil. A scale can be placed under the pointer indicator to denote the magnitude of the current flow.

The scales used to indicate current magnitudes are of two types, as shown in Fig. 5–2: (a) those with zero reference at the extreme left, and (b) those with zero reference in the center.

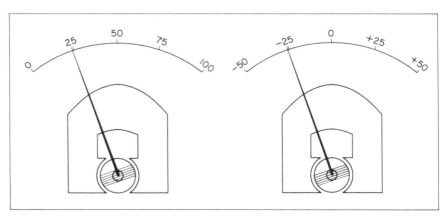

figure 5–2
(a) zero reference to the left; (b) zero reference at center

The movement is adjusted to indicate zero meter deflection when the current through the coil is zero. If the zero reference is at the far left, then the meter will indicate only when current is applied in the proper direction through the coil. Reversal of current direction would cause the pointer to rotate to the left of the zero reference, which could cause damage to the meter. Consequently, it is extremely important to observe the current meter polarity when one is using the meter. All dc meters have their terminals marked $+$ and $-$, and they must be connected so that the current flows into the $+$ terminal and leaves by the $-$ terminal.

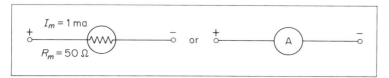

figure 5–3
real D'Arsonval meter

It is possible to construct zero center meters that can measure currents in either direction. The direction of rotation of the dial pointer depends on the current direction in the coil.

Meter movements are rated in terms of current and resistance. A typical rating for a meter is 1 ma and 50 Ω internal or meter resistance. This means that 1 ma is the current required for full-scale deflection, and the internal resistance of the movement itself is 50 ohms. The method used to represent the movement and its specifications is symbolized by the circuit shown in Fig. 5–3.

ammeters

A current greater than full scale must never be forced through a meter. An ammeter is a basic instrument used to measure current. The current

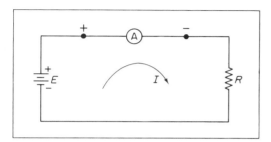

figure 5–4
measuring current with an ammeter

through any circuit resistor can be measured by placing the proper ammeter in *series* with the element, as shown in Fig. 5–4.

It is evident that the maximum current that can be measured is equal to the full-scale deflection capability of the meter. Larger currents can be measured by incorporating additional circuitry. This additional circuitry usually consists of an external resistor placed across the meter movement, as shown in Fig. 5–5. This additional resistor is called a shunt resistor and is symbolized by R_{sh}. A portion of the total input current will flow through the meter movement and the rest of the input current is *shunted* through the resistor R_{sh}.

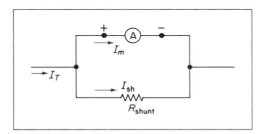

figure 5–5
constructing an ammeter by using a shunt resistor

Note that the voltage drop across the shunt resistor and the meter movement is constant and equal to

$$(I_{shunt})(R_{shunt}) = V_{meter} = V_{shunt}$$

Consequently, the current through the shunt resistor is

$$I_{sh} = I_m \frac{R_m}{R_{sh}}$$

and

$$I_T = I_{sh} + I_m$$

The shunt current can be eliminated and the total input current used to determine the shunt resistance R_{sh}. Thus,

$$R_{sh} = \frac{I_m R_m}{I_T - I_m}$$

Some illustrative problems will demonstrate the theory.

sample problem

Find the proper shunt resistor R_{sh} to construct a 1-ampere full-scale ammeter using a D'Arsonval meter movement having I_m equal to 1 ma full-scale deflection and R_m equal to 50 Ω.

solution

Calculate R_{sh}.

$$R_{sh} = \frac{I_m R_m}{I_T - I_m} \, \Omega$$

$$R_{sh} = \frac{10^{-3} \times 50}{1 - 0.001}$$

$$R_{sh} \approx 0.050 \, \Omega$$

The resultant meter is shown in Fig. 5–6.

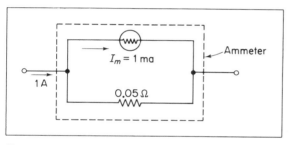

figure 5–6
one-ampere ammeter

One method of constructing a multirange ammeter is shown in Fig. 5–7, where the rotary switch determines the proper R_{sh} to be used. The shunt resistors are calculated by using the formula

$$R_{sh} = \frac{I_m R_m}{I_T - I_m}$$

Thus

$$R_{sh_1} = \frac{10^{-3} \times 50}{9 \times 10^{-3}} = 5.56 \, \Omega$$

$$R_{sh_2} = \frac{10^{-3} \times 50}{99 \times 10^{-3}} = 0.505 \, \Omega$$

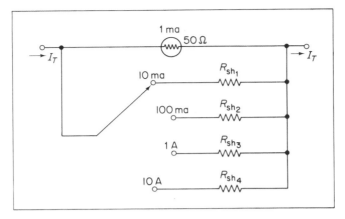

figure 5–7
multirange ammeter

$$R_{sh_3} = \frac{10^{-3} \times 50}{999 \times 10^{-3}} = 0.05\ \Omega$$

$$R_{sh_4} = \frac{10^{-3} \times 50}{9.999} = 0.005\ \Omega$$

voltmeters

The D'Arsonval meter movement can be utilized as a voltmeter by inserting a resistor in series with the meter. The basic circuit of the voltmeter is shown in Fig. 5–8. The resistor R_s is adjusted to limit the current flow to a maximum of 1 ma depending on the applied voltage.

The purpose of a voltmeter is to measure the voltage *across* an element; consequently, the meter is always connected in parallel with the element.

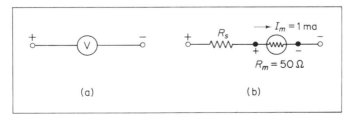

figure 5–8
a basic voltmeter: (a) symbol; (b) construction

For meter indication to register properly, the polarities of the voltage being measured must correspond to the meter terminals.

Consider the basic voltmeter shown in Fig. 5–8(b). The equation for the voltmeter circuit is

$$V = I_m R_s + I_m R_m$$

The equation can be solved for the required series meter resistance by rearranging the formula accordingly. Thus,

$$R_s = \frac{V - I_m R_m}{I_m} = \frac{V}{I_m} - R_m$$

The resultant equation indicates the value of the series multiplier resistors evaluated by taking the ratio of the desired meter voltage to the current full-scale deflection and subtracting the meter resistance from the ratio. An example will demonstrate the theory.

sample problem

Determine the series resistor R_s to construct a 1.5 voltmeter full-scale deflection using a basic meter movement having $I_m = 1$ ma and $R_m = 50\ \Omega$.

solution

The series multiplier resistance is determined by

$$R_s = \frac{V}{I_m} - R_m$$

$$R_s = \frac{1.5}{10^{-3}} - 50$$

$$R_s = 1450\ \Omega$$

To increase the meter range to 10 volts, the series multiplier resistor R_s should be

$$R_s = \frac{10}{10^{-3}} - 50$$

$$R_s = 9950\ \Omega$$

The basic design of a voltmeter can readily be extended to a multiple-range voltmeter by means of a range selectro switch and the proper multiplier resistors. The voltmeter multipliers must be arranged as shown in Fig. 5–9 or in Fig. 5–10.

Since the practical circuit is obviously the better circuit, the analysis of the multirange voltmeter will be based on this circuit. If the range switch is in position 1, the circuit equation is

$$V_{in} = I_m(R_1 + R_m)$$

and

$$R_1 = \frac{V_{in}}{I_m} - R_m$$

Move the range switch to position 2. The circuit equation is

$$R_2 = \frac{V_{in}}{I_m} - R_1 - R_m$$

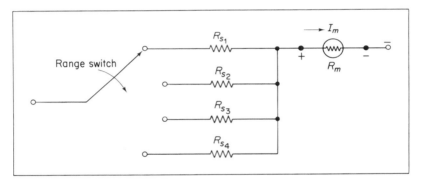

figure 5–9
multiple-range voltmeter

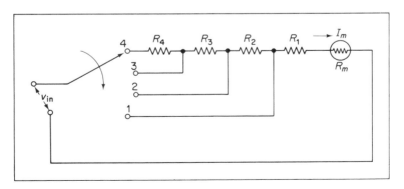

figure 5–10
practical multirange voltmeter

Rotating the rotary switch to position 3 yields the following equation.

$$V_{in} = I_m(R_1 + R_m + R_2 + R_3)$$

and

$$R_3 = \frac{V_{in}}{I_m} - R_m - R_1 - R_2$$

Continue the motion of the switch to position 4. The circuit analysis is

$$V_{in} = (R_m + R_1 + R_2 + R_3 + R_4)I_m$$

and

$$R_4 = \frac{V_{in}}{I_m} - R_m - R_1 - R_2 - R_3$$

An illustrative problem will demonstrate the theory.

sample problem

Using a basic meter movement having an I_m equal to 1 ma and an R_m equal to 50 ohms, design a multiple-range voltmeter having 5-v, 10-v, 50-v, 100-v, and 500-v ranges.

solution

step 1: Refer to Fig. 5–11. For the 5-v range, the value of R_1 is (switch in position 1)

$$R_1 = \frac{V_{in}}{I_m} - R_m$$

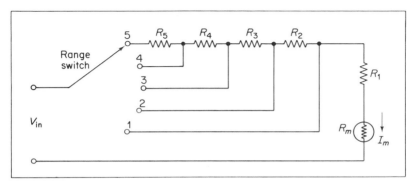

figure 5–11

$$R_1 = \frac{5}{10^{-3}} - 50$$

$$R_1 = 4950 \ \Omega$$

step 2:   For the 10-v range, the value of R_2 is calculated by

$$R_2 = \frac{V_{in}}{I_m} - R_1 - R_m$$

$$R_2 = \frac{10}{10^{-3}} - 4950 - 50$$

$$R_2 = 5 \ k\Omega$$

step 3:   For the 50-v range, the value of R_3 is

$$R_3 = \frac{V_{in}}{I_m} - R_1 - R_2 - R_m$$

$$R_3 = \frac{50}{10^{-3}} - 4950 - 5000 - 50$$

$$R_3 = 40 \ k\Omega$$

step 4:   For the 100-v range, the value of R_4 is

$$R_4 = \frac{V_{in}}{I_m} - R_1 - R_2 - R_3 - R_m$$

$$R_4 = \frac{100}{10^{-3}} - 4950 - 5000 - 40,000 - 50$$

$$R_4 = 50 \ k\Omega$$

step 5:   For the 500-v range, the value of R_5 is calculated by

$$R_5 = \frac{V_{in}}{I_m} - R_1 - R_2 - R_3 - R_4 - R_m$$

$$R_5 = \frac{500}{10^{-3}} - 4950 - 5000 - 40,000 - 50,000 - 50$$

$$R_5 = 400 \ k\Omega$$

voltage sensitivity

The input resistance of a voltmeter is relatively important, since it determines the "loading effect" of the meter on the circuit being measured. The input resistance of an ideal voltmeter should be infinite.

A practical voltmeter has some finite resistance that will take some current from the circuit under investigation. The total resistance of the voltmeter R_t on any voltmeter scale can readily be determined as equal to the ratio of the full-scale voltmeter input divided by the full-scale deflection meter current. Thus,

$$R_t = \frac{V_{in}}{RI_m}$$

Thus, when

$$I_m = 1 \text{ ma} \quad \text{and} \quad \begin{cases} V_{in} = 100 \text{ v} \\ V_{in} = 10 \text{ v} \end{cases}$$

then

$$R_t = \frac{100}{10^{-3}} = 100 \text{ k}\Omega$$

and

$$R_t = \frac{10}{10^{-3}} = 10 \text{ k}\Omega$$

It is evident that as the input voltage increases, the voltmeter resistance increases. Note that the meter resistance R_t is directly proportional to the input voltage V_{in}.

The voltmeter sensitivity is defined as the ohm per volt rating of the meter. The symbol for the voltage sensitivity is the Greek letter nu (ν). Thus,

$$\nu = \frac{\text{ohm } (\Omega)}{\text{volt } (V)}$$

The sensitivity ν is a constant for any meter and is usually printed on the face of the meter.

---

sample problem 1

What is the sensitivity of a meter having an I_m equal to 200 μA ?

---

solution

$$\nu = \frac{\text{ohms}}{\text{volt}} = \frac{1}{I_m}$$

$$\nu = \frac{1}{200 \times 10^{-6}}$$

$$\nu = 5 \text{ k}\Omega$$

solution

$$I_m = \frac{1}{v}$$

$$I_m = \frac{1}{100 \times 10^3} = 10 \ \mu A$$

ohmmeters

An ohmmeter is a device used to measure resistance. The basic circuit of an ohmmeter is shown in Fig. 5–12. The resistor R_x represents the unknown

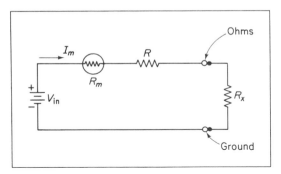

figure 5–12
a basic ohmmeter

resistance and is connected across the output terminals marked "ohms" and "ground," respectively.

analysis

Assume that R_x is removed and R_m is approximately equal to zero. Then shorting the output terminals is essentially permitting R_x to be zero. Then

$$I_m = \frac{V_{in}}{R}$$

The values of V and R, respectively, are such as to cause full-scale deflection of the meter under short-circuit conditions.

Remove the unknown resistor R_x from the output terminals so that the terminals are open-circuited. The meter current is zero, corresponding to R_x's being equal to infinity.

Note that for any value of R_x between the limits of zero and infinity, the circuit equation is

$$R_x = \frac{V_{\text{in}}}{I_m} - R$$

An illustrative problem will demonstrate the theory.

---

sample problem

A circuit for a single-scale ohmmeter is shown in Fig. 5–13. The circuit must have a 10-kΩ center scale.

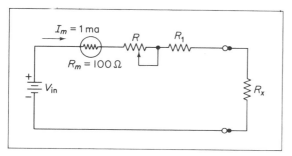

figure 5–13
an ohmmeter

solution

step 1: Since the center-scale value is to be 10 kΩ, then $R_1 + R + R_m$ must equal 10 kΩ. R_1 is chosen as approximately half of the center-scale value, or 5 kΩ. Thus, R would comprise a 10-kΩ potentiometer for proper adjustment. Then

$$R_T = R_1 + R + R_m = 10 \text{ k}\Omega$$

step 2: Calculate V_{in}.

$$V_{\text{in}} = I_m R_T$$
$$V_{\text{in}} = 10^{-3} \times 10^4 = 10 \text{ v}$$

step 3: Calculate and tabulate I_m for R_x equal to 0.1 R_T to 10 R_T. The equation used is

$$I_m = \frac{V_{in}}{R_T + R_x}$$

I_m in ma	R_x
0.909	1 kΩ
0.833	2 kΩ
0.667	5 kΩ
0.588	7 kΩ
0.500	10 kΩ
0.333	20 kΩ
0.167	50 kΩ
0.125	70 kΩ
0.0909	100 kΩ

step 4: The ohmmeter scale appears as shown in Fig. 5–14.

Wheatstone bridge

The Wheatstone bridge shown in Fig. 5–15 is a method used accurately to measure the unknown resistance R_x. An extremely sensitive zero-center galvonometer is placed between nodes A and B. The resistors R_1 and R_2 are standard resistors. R_3 is adjusted until the galvonometer shows zero deflection, which means that the bridge is balanced. When the bridge circuit is balanced, there is no current through the galvonometer, because the voltage from A to C is exactly equal to the voltage from B to C. The voltage equations for a balanced bridge are

$$I_1 R_1 = I_2 R_x$$

and

$$I_1 R_2 = I_2 R_3$$

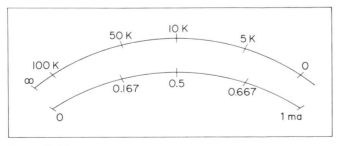

figure 5–14
ohmmeter

Dividing the two equations yields

$$\frac{I_1 R_1}{I_1 R_2} = \frac{I_2 R_x}{I_2 R_3}$$

and

$$R_x = R_3 \frac{R_1}{R_2}$$

It should be noted that the battery voltage V_{in} does not appear in the balance equations. A specific advantage of this system over the basic ohmmeter is that the circuit balance is independent of the variations in the source voltage.

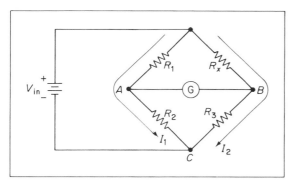

figure 5–15
Wheatstone bridge circuit

For reasonable accuracy in determining R_x, precision resistors should be used for R_1, R_2, and R_3. Values of R_1 and R_2 should be close in value to each other for reasonable accuracy. R_3 should then be reasonably close to R_x for extreme accuracy in the bridge measurement.

An illustrative problem will demonstrate the theory.

sample problem

The Wheatstone bridge circuit shown in Fig. 5–15 has the following given data.

$$R_1 = 50 \text{ k}\Omega, \qquad R_2 = 100 \text{ k}\Omega, \qquad R_3 = 50 \text{ k}\Omega$$

Find R_x.

solution

The required equation is

$$R_x = R_3 \frac{R_1}{R_2} = 50 \frac{50}{100} 10^3$$

$$R_x = 25 \text{ k}\Omega$$

problems

1. What is the terminal voltage across a D'Arsonval meter having an R_m of 50 Ω and an I_m of 100 ma?

2. What is the terminal voltage across a D'Arsonval meter having an R_m of 100 Ω and an I_m of 400 μA?

3. If the voltage across a D'Arsonval meter is 50 mv when I_m is 1 ma, what must be the meter resistance?

4. What is the meter current if the terminal voltage across the meter of 100 Ω is 100 mv?

5. A D'Arsonval meter has an $R_m = 50 \Omega$ and an $I_m = 500 \mu$A. Determine the value of the shunt resistor required to construct a 2-ma meter.

6. A meter movement has a resistance of 50 Ω and a full-scale deflection of 2 ma. Determine the value of the series multiplier to convert the movement to read 300 v full-scale deflection.

7. A meter movement has an R_m of 2 kΩ and an I_m of 50 μA. Find the value of a series multiplier to convert the meter to read 2500 volts full-scale deflection.

8. A meter movement has an R_m of 100 Ω and an I_m of 1 ma. Determine the series multiplier necessary to convert the movement to indicate 5 v, 50 v, and 500 v.

9. A meter movement has an R_m of 50 Ω and an I_m of 50 μA. Determine the series multiplier required to convert the meter to 3 v, 30 v, and 300 v.

10. A voltmeter is constructed that uses a meter movement having an R_m of 100 Ω and an I_m of 1 ma. The series multiplier used is a 6-kΩ resistor. What value resistor must be placed in parallel with the resistor to make the meter deflect full scale when the applied voltage is 4 v?

11. A 250-v voltmeter having a resistance of 10 kΩ is used to measure an unknown voltage. The meter goes off scale. When a 5 kΩ resistor is placed in series with the meter, the indicator reads 200 v. What is the unknown voltage?

12. What is the full-scale deflection current of a meter having a sensitivity of 25 kΩ/volt?

13. Using a basic meter movement with an I_m of 500 μA and an R_m of 50 Ω, design a simple ohmmeter having a 1-kΩ center scale.

14. A 1-ma, 1000-Ω meter movement is used with a 45-volt battery in series in an ohmmeter circuit. What is the half-scale deflection?

15. Given the voltmeter circuit shown, find R_2 and R_3.

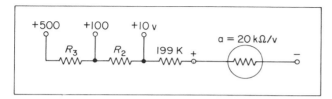

16. If the sensitivity of the meter movement in Problem 15 were 10 kΩ/volt, determine the value of R_1, R_2, and R_3. $R_m = 1$ kΩ.

17. A 1-ma, 10-kΩ meter movement is used to construct an ohmmeter with a 100-kΩ center-scale deflection. Determine system elements.

18. In the Wheatstone bridge circuit shown in Fig. 5–15, the data given are
$$R_1 = 32 \ \Omega, \qquad R_2 = 42 \ \Omega, \qquad R_3 = 34 \ \Omega, \qquad R_4 = 36 \ \Omega$$
Is the bridge balanced?

19. In the Wheatstone bridge circuit shown in Fig. 5–15, the data given are
$$R_1 = 10 \ \text{k}\Omega, \qquad R_2 = 5 \ \text{k}\Omega$$
R_3 is adjusted to 675 Ω for bridge balance. Determine the value of R_x.

20. In the Wheatstone bridge circuit shown in Fig. 5–15, the data given are
$$R_1 = 100 \ \text{k}\Omega, \qquad R_2 = 25 \ \text{k}\Omega$$
R_3 is adjusted to 21.5 kΩ for zero balance. Determine the value of R_x.

6
network theorems

π *to T network conversion*

The characteristics of dc networks enabled us to understand the operation of simple circuits. Note that a complete series parallel combination was replaced by a single resistor that was electrically equivalent to the complete network. This technique of equivalency is the underlying principle on which many of our network theorems are based.

A useful transformation that replaces one network by an equivalent circuit is called the "π to T" conversion. Consider the four-terminal network shown in Fig. 6–1. For the two networks to be equivalent, the resistances at each set of terminals must be identical. The following nomenclature will be used to denote various electrical measurements.

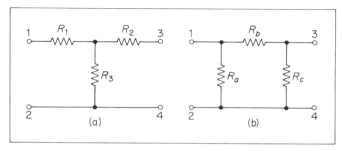

figure 6–1
(a) *T* network; (b) π network

R_{12} = input resistance between terminals 1 and 2

R_{13} = input resistance between terminals 1 and 3

R_{23} = input resistance between terminals 2 and 3

Since there is a common connection between terminals 2 and 4, a fourth measurement is not required.

Consider the situation whereby the elements of the π network are known and the problem is to determine the elements of an equivalent T network. Then

π	T	
$R_{12} = \dfrac{R_a(R_b + R_c)}{R_a + R_b + R_c}$	$R_{12} = R_1 + R_3$	(6–1)
$R_{13} = \dfrac{R_b(R_a + R_c)}{R_a + R_b + R_c}$	$R_{13} = R_1 + R_2$	(6–2)
$R_{23} = \dfrac{R_c(R_a + R_b)}{R_a + R_b + R_c}$	$R_{23} = R_2 + R_3$	(6–3)

For the two circuits to be equivalent, the resistances at each set of terminals must be identical. Thus, $R_{12_\pi} = R_{12_T}$, etc. To determine one element of the equivalent T, add R_{12} and R_{13} and subtract R_{23} from the resultant. Thus,

$$\begin{array}{cc} T & \pi \\ 2R_1 = & \dfrac{2R_aR_b}{R_a + R_b + R_c} \end{array} \tag{6–4}$$

It is evident that the arm R_1 is equal to

$$R_1 = \frac{R_aR_b}{R_a + R_b + R_c} \tag{6–5}$$

To determine R_2, add R_{13} and R_{23} and subtract R_{12} from the resultant. Then

$$R_2 = \frac{R_bR_c}{R_a + R_b + R_c} \tag{6–6}$$

The third element R_3 is found by adding R_{12} to R_{23} and subtracting R_{13} from the resultant. This yields

$$R_3 = \frac{R_a R_c}{R_a + R_b + R_c} \qquad (6\text{-}7)$$

Physically, this analysis shows that the π network can be replaced by an equivalent T without any variation in the circuit characteristic. Note that in the four-terminal network, the T network replaces the π as shown in Fig. 6-2.

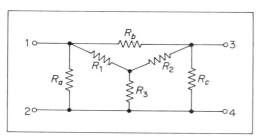

figure 6-2
replacing the π by an equivalent T

An example will demonstrate the theory.

sample problem

The arms of a π network have the following values.
$R_a = 625\ \Omega,\qquad R_b = 250\ \Omega,\qquad R_c = 375\ \Omega$
Find the equivalent T network.

solution

step 1:   Calculate R_1.

$$R_1 = \frac{R_a R_b}{R_a + R_b + R_c}$$

$$R_1 = \frac{625 \times 250}{1250} = 125\ \Omega$$

step 2:   Calculate R_2.

$$R_2 = \frac{R_b R_c}{R_a + R_b + R_c} = \frac{250 \times 375}{1250}$$

$$R_2 = 75\ \Omega$$

step 3 : Calculate R_3.

$$R_3 = \frac{R_a R_c}{R_a + R_b + R_c} = \frac{625 \times 375}{1250}$$

$$R_3 = 137.5 \ \Omega$$

Investigation of the network conversion formulas indicates the following: *To determine any arm of the T network, multiply the two arms adjacent to the desired arm and divide by the sum of the π elements.* This product over sum rule can be applied by reference to Fig. 6–3.

$$R_1 = \frac{\text{product of the sides adjacent}}{\text{sum of the elements in the } \pi \text{ network}}$$

or

$$R_1 = \frac{R_a R_b}{R_a + R_b + R_c}$$

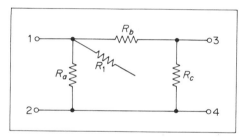

figure 6–3
demonstrating product over sum rule

When each arm of the π network is equal, then under these conditions, $R_1 = R_2 = R_3 = R_a/3$. If each arm of the π network is equal to 15 ohms, then the equivalent T network has each arm to 5 ohms. This is shown in Fig. 6–4.

T to π conversion

The same general approach may be used to convert from the T to the π network. It has been established that

$$R_1 = \frac{R_a R_b}{R_a + R_b + R_c}, \qquad R_2 = \frac{R_b R_c}{R_a + R_b + R_c}, \qquad R_3 = \frac{R_a R_c}{R_a + R_b + R_c}$$

If R_1 is divided by R_2, then

$$\frac{R_1}{R_2} = \frac{R_a}{R_c} \quad \text{and} \quad \frac{R_1}{R_3} = \frac{R_b}{R_c}$$

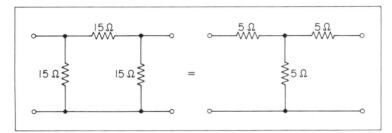

figure 6–4
special case of π to T conversion

Substituting these values into the equation for R_1 yields

$$R_1 = \frac{\left(\dfrac{R_1}{R_2} R_c\right)\left(\dfrac{R_1}{R_3} R_c\right)}{\left(\dfrac{R_1}{R_2}\right) R_c + \left(\dfrac{R_1}{R_3}\right) R_c + R_c}$$

Solving this equation results in

$$R_c = \frac{R_1 R_3 + R_1 R_2 + R_2 R_3}{R_1} \tag{6-8}$$

Using the same technique, we find that the resultant values for R_a and R_b are

$$R_a = \frac{R_1 R_2 + R_1 R_3 + R_2 R_3}{R_2} \tag{6-9}$$

$$R_b = \frac{R_1 R_2 + R_1 R_3 + R_2 R_3}{R_3} \tag{6-10}$$

Investigation of these network formulas permits us to formulate a rule: The arms of a π network that is equivalent to the T network are equal to the *sum of the products of any two arms divided by the open arm*. This can be demonstrated by reference to Fig. 6–5. Note that

$$R_a = \frac{R_1 R_2 + R_1 R_3 + R_2 R_3}{R_2}$$

An illustrative problem will demonstrate the theory.

sample problem

The arms of a T network are given, respectively, by
$R_1 = 240\ \Omega$, $R_2 = 960\ \Omega$, $R_3 = 600\ \Omega$
Determine the equivalent π network.

solution

step 1:   Calculate the sum of the products.

$$R_1 R_2 = 240 \times 960 = 230.4 \times 10^3$$
$$R_1 R_3 = 240 \times 600 = 144 \times 10^3$$
$$R_2 R_3 = 960 \times 600 = 576 \times 10^3$$

The total sum is 950.4×10^3.

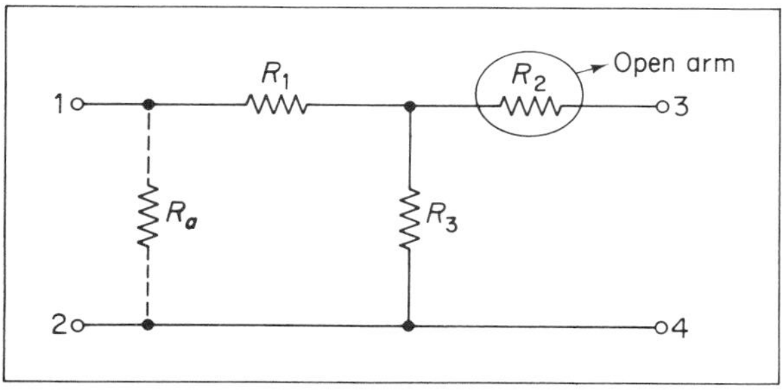

figure 6–5
demonstrating T to π conversion rule

step 2:   Calculate R_a.

Let the sum of the products be designated by N. Thus,

$$R_a = \frac{N}{R_2} = \frac{950.4 \times 10^3}{960}$$

$$R_a = 990\ \Omega$$

step 3:   Calculate R_b.

$$R_b = \frac{N}{R_3} = \frac{950.4 \times 10^3}{600}$$

$$R_b = 1584\ \Omega$$

step 4:   Calculate R_c.

$$R_c = \frac{N}{R_1} = \frac{950.4 \times 10^3}{240}$$

$$R_c = 3960 \ \Omega$$

superposition theorem

In many electrical networks, the circuit may have more than one source of potential applied. The total current that flows in any branch of the circuit is a function of all the applied voltage sources. For example, consider the circuit shown in Fig. 6–6. Assume that the problem is to find I_3. The solution to this problem is not a simple one, except that we can state that the current I_3 must be some function of E_1 and E_2.

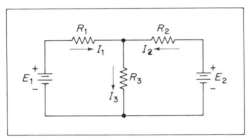

figure 6–6
circuit for the superposition theorem

The superposition theorem can prove useful in problems of this type. The superposition theorem states that *the current flowing through any element of a linear network is the algebraic sum of the currents produced by each source individually with all other sources removed and replaced by their internal resistances.*

problem

Consider the network shown in Fig. 6–7. Since there are two sources, two circuits are required to demonstrate the theorem.

In the circuit shown in Fig. 6–8 the total resistance across the source is

$$R_T = 20 + \frac{90 \times 60}{150}$$

$$R_T = 56 \ \Omega$$

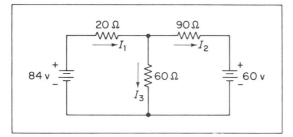

figure 6–7
circuit for problem

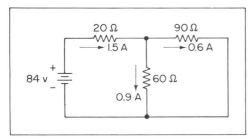

figure 6–8
current components due to 84-volt source

The source current is

$$I_T = \tfrac{84}{56} = 1.5 \text{ amperes}$$

The current through the 60-ohm resistor is using the current divider theorem equal to

$$I_{60\Omega} = 1.5 \frac{90}{90 + 60}$$

$$I_{60\Omega} = 0.9 \text{ ampere}$$

The current through the 90-ohm resistor is equal to

$$I_{90\Omega} = I_T - I_{60\Omega} = 1.5 - 0.9 = 0.6 \text{ ampere}$$

These currents and their respective directions are shown in Fig. 6–9.
Removing the other source and inserting the 60-volt source yields the circuit shown in Fig. 6–9. The total resistance is

$$R_T = 90 + \left(\frac{20 \times 60}{20 + 60}\right) = 105 \text{ ohms}$$

The total current is

$$I_T = \tfrac{60}{105} = 0.57 \text{ ampere}$$

Using the current divider rule, we find that the current through the 60-ohm resistor is

$$I_{60\Omega} = 0.57 \left(\frac{20}{20+60} \right) = 0.1425 \text{ ampere}$$

The current through the 20-ohm resistor is

$$I_{20\Omega} = 0.57 - 0.1425 = 0.4275 \text{ ampere}$$

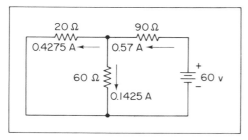

figure 6–9
current components due to 60-volt source

The two sets of current values are algebraically additive. The direction of the current arrows shown in Fig. 6–9 is assumed to be the positive direction, so we have the following resultant values of current.

$$I_1 = 1.5 - 0.4275 = 1.0725 \text{ amperes}$$

$$I_2 = 0.6 - 0.57 = 0.03 \text{ ampere}$$

$$I_3 = 0.9 + 0.1425 = 1.0425 \text{ amperes}$$

Note that the total current I_1 must be equal to the sum of the currents of I_2 and I_3. This can be used as a means of checking the solution. In this case, the results check, since

$$I_1 = 1.0725 = 1.0425 + 0.03$$

sample problem

The following circuit is given. Determine the branch currents, using the superposition theorem,

where $E_1 = 20$ v $R_1 = 2$ kΩ
        $E_2 = 60$ v $R_2 = 4$ kΩ
        $E_3 = 80$ v $R_3 = 12$ kΩ
                              $R_L = 6$ kΩ

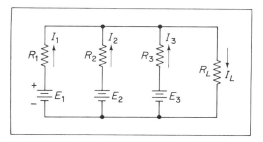

solution

step 1:   Remove the various sources and determine the current components due to E_1 only. Thus, the resultant circuit is as shown in the figure.

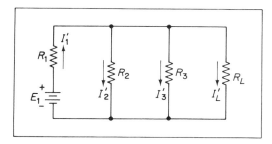

step 2:   Calculate the total resistance.

$$R_T = R_1 + \cfrac{1}{\cfrac{1}{R_2} + \cfrac{1}{R_3} + \cfrac{1}{R_L}}$$

$R_T = 4$ kΩ

step 3:   Calculate the source current.

$$I_1' = \frac{E_1}{R_T} = \frac{20}{4 \times 10^3}$$

$I_1' = 5$ ma

step 4: Calculate the remaining current components. The voltage across R_1 is

$$E_{R_1} = I_1' R_1 = 5 \times 10^{-3} \times 2 \times 10^3 = 10 \text{ v}$$

The voltage E_p across the parallel combination is

$$E_p = E_1 - E_{R_1} = 20 - 10 = 10 \text{ v}$$

The current components are

$$I_2' = \frac{E_p}{R_2} = \frac{10}{4 \times 10^3} = 2.5 \text{ ma}$$

$$I_3' = \frac{E_p}{R_3} = \frac{10}{12 \times 10^3} = 0.833 \text{ ma}$$

$$I_L' = \frac{E_p}{R_L} = \frac{10}{6 \times 10^3} = 1.667 \text{ ma}$$

step 5: Remove all sources except E_2. Determine the total resistance.

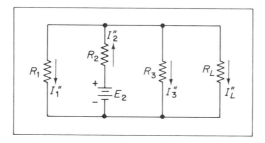

$$R_T = R_2 + \cfrac{1}{\cfrac{1}{R_1} + \cfrac{1}{R_3} + \cfrac{1}{R_L}}$$

$$R_t = 5.33 \text{ k}\Omega$$

step 6: The source current is

$$I_2'' = \frac{E_2}{R_T} = \frac{60}{5.33 \times 10^3}$$

$$I'' = 11.25 \text{ ma}$$

step 7: Determine the other current components. The voltage drop across R_2 is

$$E_{R_2} = I'' R_2 = 11.25 \times 4 = 45 \text{ v}$$

The voltage E_p across the parallel combination is

$$E_p = E_2 - E_{R_2} = 60 - 45 = 15 \text{ v}$$

The current components are

$$I''_1 = \frac{E_p}{R_1} = \frac{15}{2 \times 10^3} = 7.5 \text{ ma}$$

$$I''_3 = \frac{E_p}{R_3} = \frac{15}{12 \times 10^3} = 1.25 \text{ ma}$$

$$I''_L = \frac{E_p}{R_L} = \frac{15}{6 \times 10^3} = 2.5 \text{ ma}$$

step 8: Remove all sources except E_3. Determine the total resistance.

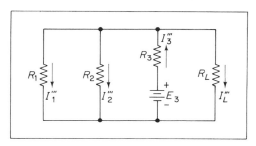

$$R_T = R_3 + \cfrac{1}{\cfrac{1}{R_1} + \cfrac{1}{R_2} + \cfrac{1}{R_L}}$$

$$R_T = 13.09 \text{ k}\Omega$$

step 9: The source current is

$$I'''_3 = \frac{E_3}{R_T} = \frac{80}{13.09 \times 10^3} = 6.11 \text{ ma}$$

step 10: Determine the current components. The voltage drop across R_3 is

$$E_{R_3} = I'''_3 R_3 = 6.11 \times 12 = 73.32 \text{ v}$$

The voltage E_p across the parallel combination is

$$E_p = E_3 - E_{R_3} = 80 - 73.32 = 6.68 \text{ v}$$

The current components are

$$I'''_1 = \frac{E_p}{R_1} = \frac{6.68}{2 \times 10^3} = 3.34 \text{ ma}$$

$$I'''_2 = \frac{E_p}{R_2} = \frac{6.68}{4 \times 10^3} = 1.67 \text{ ma}$$

$$I'''_L = \frac{E_p}{R_L} = \frac{6.68}{6 \times 10^3} = 1.11 \text{ ma}$$

step 11: The total branch currents can now be evaluated. Assume that the current arrows from the original circuit have a positive direction. Then

$$I_1 = I_1' - I_1'' - I_1''' = (5 - 7.5 - 3.34) \text{ ma}$$

$$I_1 = -5.84 \text{ ma}$$

The negative sign indicates that the original current direction was assumed incorrectly and should be reversed.

$$I_2 = I_2'' - I_2' - I_2''' = (11.25 - 2.5 - 1.67) \text{ ma}$$

$$I_2 = 7.08 \text{ ma}$$

$$I_3 = I_3''' - I_3' - I_3'' = (6.11 - 0.833 - 1.25) \text{ ma}$$

$$I_3 = 4.027 \text{ ma}$$

$$I_L = I_L' + I_L'' + I_L''' = (1.667 + 2.5 + 1.11) \text{ ma}$$

$$I_L = 5.277 \text{ ma}$$

step 12: The resultant circuit is as shown in the figure.

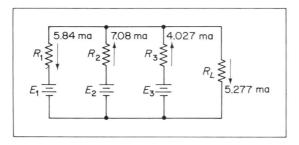

Thevenin's theorem

The concept of replacing a complex network by a single equivalent resistor can be further extended to include multisource networks. A theorem that utilizes this technique was developed by Thevenin and is called Thevenin's theorem. This theorem in its essence states: _Any linear network that is energized by one or more sources of energy can be replaced by an equivalent resistance in series with one energy source._

Since one picture is worth a million words, let us examine the pictorial representation of the theorem as shown in Fig. 6–10.

Open-circuit the load terminals at points _a–b_. Measure the voltage across terminals _a–b_. This voltage is called E_{oc}, or the open-circuit voltage. Remove the sources and replace each source by its internal resistance.

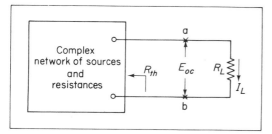

figure 6–10
pictorial representation of Thevenin's theorem

Measure the resistance looking back into the circuit from terminals *a–b*. This resistance is called R_{th}. The entire complex network can be removed and replaced by an equivalent network, as shown in Fig. 6–11. The current through the load is

$$I_L = \frac{E_{oc}}{R_{th} + R_L}$$

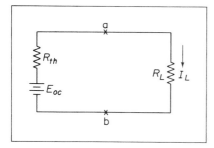

figure 6–11
Thevenin's equivalent circuit

An illustrative problem will demonstrate the theory.

---

sample problem

Find the equivalent circuit and the current in R_L by Thevenin's theorem for the circuit shown,

where $E = 100$ v $R_d = 35$ kΩ
        $R_a = 2$ kΩ $R_e = 13.33$ kΩ
        $R_b = 15$ kΩ $R_L = 11.5$ kΩ
        $R_c = 25$ kΩ

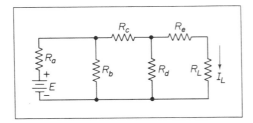

solution

step 1: Convert the inner π to an equivalent T.

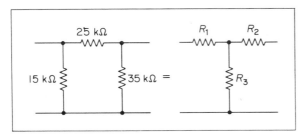

$$R_1 = \frac{15 \times 25}{75} \text{ k}\Omega$$

$$R_1 = 5 \text{ k}\Omega$$

$$R_2 = \frac{25 \times 35}{75} = 11.67 \text{ k}\Omega$$

$$R_3 = \frac{15 \times 35}{75} = 7 \text{ k}\Omega$$

The new circuit is as shown in the figure.

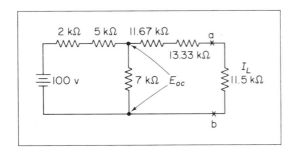

step 2: Calculate E_{oc}. Since terminals a–b are open, then E_{oc} is the voltage across the 7-kilohm resistor. Thus,

$$E_{oc} = \frac{100 \times 7 \times 10^3}{(7+7)10^3} = 50 \text{ v}$$

step 3: Calculate R_{th}. The 100-volt source is removed and replaced by a short circuit. Then

$$R_{th} = \left(25 + \frac{7 \times 7}{7+7}\right) 10^3$$

$$R_{th} = 28.5 \text{ k}\Omega$$

step 4: The basic equivalent circuit becomes as shown in the figure.

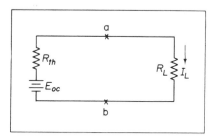

step 5: Calculate I_L.

$$I_L = \frac{E_{oc}}{R_{th} + R_L} = \frac{50}{40 \times 10^3}$$

$$I_L = 1.25 \text{ ma}$$

Norton's theorem

Norton's theorem states that *any linear network that is energized by one or more sources of energy can be replaced by a single constant current source in parallel with an equivalent resistance.*

The pictorial representation of this theorem is shown in Fig. 6–12. Note that Thevenin's theorem is considered a voltage equivalent circuit. The same technique can be used to establish Norton's theorem as a current equivalent circuit.

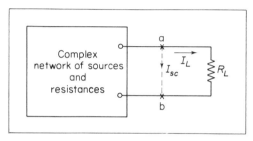

figure 6–12
pictorial representation of Norton's theorem

The complex network of sources and resistances can be replaced by a Thevenin's equivalent circuit, as shown in Fig. 6–13. The application of Norton's theorem to the circuit of Fig. 6–13 assumes that the load is removed. Place a short circuit between terminals a–b. Measure the short circuit current.

$$I_{sc} = \frac{E_{oc}}{R_{th}}$$

Remove the short circuit and E_{oc}. Replace E_{oc} by a short circuit. Measure the resistance looking into the circuit at terminals a–b.

$$R_{a-b} = R_{th}$$

The resultant equivalent circuit is shown in Fig. 6–14.

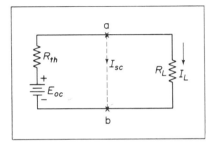

figure 6–13
Thevenin's equivalent circuit

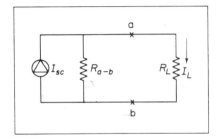

figure 6–14
Norton's equivalent circuit

In most circuit applications, the usual procedure is to Thevenize the circuit and then convert to a Norton's equivalent circuit if it is more convenient to problem solution.

Millman's theorem

A theorem that utilizes both Norton's and Thevenin's procedures in one network is known as Millman's theorem. Millman's theorem is quite useful in the field of computers and deals with multisource problems. Consider the multisource network shown in Fig. 6–15.

Specifically, Millman's theorem states that the voltage E_L is equal to

$$E_L = \frac{+\dfrac{E_1}{R_1} + \dfrac{E_2}{R_2} - \dfrac{E_3}{R_3} + \dfrac{0}{R_L}}{\dfrac{1}{R_1} + \dfrac{1}{R_2} + \dfrac{1}{R_3} + \dfrac{1}{R_L}}$$

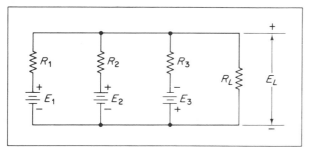

figure 6–15
circuit for Millman's theorem

It is evident that the numerator comprises every branch source divided by the branch resistance with the polarity dependent on the source voltage. The denominator contains the reciprocal of each branch resistance. An illustrative problem will demonstrate the theory.

sample problem

In the given circuit, find E_L, using Millman's theorem.

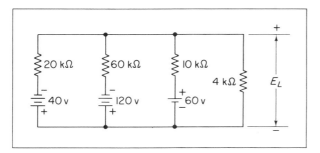

solution

$$E_L = \frac{(-\frac{40}{20} - \frac{120}{60} + \frac{60}{10})10^{-3}}{(\frac{1}{20} + \frac{1}{60} + \frac{1}{10} + \frac{1}{4})10^{-3}}$$

$$E_L = 4.8 \text{ v}$$

reciprocity theorem

Another network theorem that is useful in network analysis is the theorem of reciprocity. Consider the circuit shown in Fig. 6–16. If a voltage

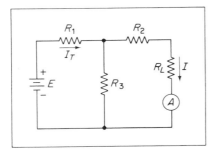

figure 6–16
circuit used for reciprocity theorem

E is applied to the input terminals, a current I measured by ammeter A will be produced in another mesh. Interchanging source voltage E and ammeter A will produce the same current I flowing through ammeter A.

Refer to Fig. 6–16. The current I that flows through the ammeter is equal to

$$I = I_T \left(\frac{R_3}{R_2 + R_3 + R_L} \right)$$

and

$$I_T = \frac{E}{R_1 + \dfrac{R_3(R_2 + R_L)}{R_2 + R_3 + R_L}}$$

Substituting and solving for I yields

$$I = \frac{ER_3}{R_1 R_3 + (R_1 + R_3)(R_2 + R_L)}$$

At this point interchange source E with ammeter A. The resultant circuit is shown in Fig. 6–17. The current I is

$$I = I_L \left(\frac{R_3}{R_1 + R_3} \right)$$

and

$$I_L = \frac{E}{R_L + R_2 + \left(\dfrac{R_1 R_3}{R_1 + R_3} \right)}$$

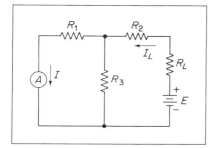

figure 6–17
circuit used for reciprocity theorem

Substituting and solving yields the following resultant equation.

$$I = \frac{ER_3}{(R_1 + R_3)(R_2 + R_L) + R_1 R_3}$$

and the theorem has been demonstrated.

maximum power transfer theorem

In communications systems, a frequent problem is to transmit or receive maximum possible power. Consider a Thevenin's equivalent circuit as shown in Fig. 6–18. Assume that R_L is variable from zero to infinity. A graph of power versus the load resistance is shown in Fig. 6–19, as R_L is varied over its entire range.

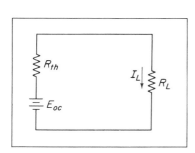

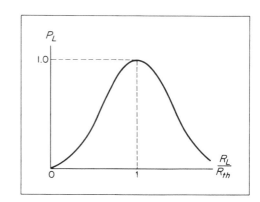

figure 6–18
circuit for maximum power
transfer theorem

figure 6–19
power vs. R_L

It is evident that the load power P_L is equal to

$$P_L = I_L^2 R_L$$

or

$$P_L = \frac{E_{oc}^2 R_L}{(R_{th} + R_L)^2}$$

If the rate of change of P_L is taken with respect to R_L the resultant value is

$$\frac{dP_L}{dR_L} = \frac{E_{oc}^2 (R_{th} + R_L)^2 - 2E_{oc}^2 (R_{th} + R_L)(R_L)}{(R_{th} + R_L)^4}$$

Setting the rate of change equal to zero and solving, we obtain

$$R_L = R_{th}$$

It is evident that for maximum power transfer the load resistance must equal the source resistance.

loop circuit analysis

The previous method of circuit analysis used the branch current technique. The more modern technique of loop circuit analysis provides a more general and systematic approach to the solution of complex network problems. Kirchhoff's voltage law may be used to determine the various loop currents within the network. Note that a restatement of the law is the following: The algebraic sum of all voltages taken in a closed loop in sequential order must equal zero.

Consider the circuit shown in Fig. 6–20. Assume current I in the clockwise direction as shown. If the current direction is incorrect, the

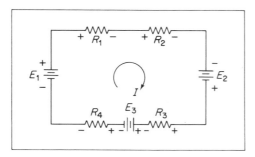

figure 6–20
series circuit for loop circuit analysis

value of current will contain a negative sign, denoting that the current direction should have been reversed. Note that the conventional system of current flow will be used; that is, the arrowhead of current enters the positive terminal of a resistor. Starting at E_1 and proceeding clockwise around the loop results in the following equation.

$$+E_1 - IR_1 - IR_2 + E_2 - IR_3 - E_3 - IR_4 = 0$$

The polarity of the voltage sources is independent of the direction of current flow. Consider a two-loop network, as shown in Fig. 6–21. Assume currents I_1 and I_2 to rotate clockwise in direction. The application of Kirchhoff's voltage law requires two equations. Applying this law to loop *abcda* results in

$$+E_1 - I_1R_1 - I_1R_2 - (I_1 - I_2)R_3 - E_2 = 0$$

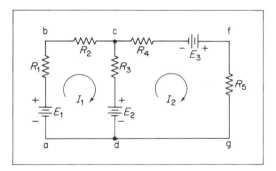

figure 6–21
two-loop network

Note that the two currents both flow through R_3. The second equation for the loop *dcfgd* is

$$+E_2 - (I_2 - I_1)R_3 - I_2R_4 + E_3 - I_2R_5 = 0$$

The two loop equations are

$$E_1 - E_2 = I_1(R_1 + R_2 + R_3) - I_2R_3$$

$$E_2 + E_3 = -I_1R_3 + I_2(R_3 + R_4 + R_5)$$

These two equations may be solved either by substitution or determinants. The procedure for systematically writing loop equations is

1. In each closed loop, assume a current direction.

2. Indicate the polarity of voltage drops or rises across each resistor in the loop.

3. Determine the algebraic sum of the voltages around each closed loop and set it equal to zero.

4. If a resistor has two or more loop currents flowing through it, the total current through the resistor is the loop current of interest plus or minus the other currents dependent on direction; plus if in the same direction, minus if opposite.

5. The polarity of all voltage sources is independent of current directions.

(If a circuit contains three or more loops, determinants or matrix methods become the desirable technique used for problem solving. A detailed discussion of determinants and matrix methods is given in the following chapter.)

An illustrative problem will demonstrate the theory.

---

sample problem

Determine I_1 and I_2, using the loop circuit method.

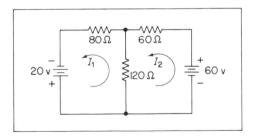

solution

step 1: The equation for loop 1 is

$$+20-(I_1-I_2)120-80I_1 = 0$$

Rewriting this equation results in

$$20 = 200I_1 - 120I_2$$

step 2: The equation for loop 2 is

$$+60-60I_2-(I_2-I_1)120 = 0$$

Rewriting this equation results in

$$60 = -120I_1 + 180I_2$$

Note the technique of writing I_1 first, then I_2, etc. This practice will simplify problem solving for future and more modern problems.

step 3: Place these two equations one below the other before proceeding with the solution of simultaneous linear equations.

$$20 = 200I_1 - 120I_2 \qquad \text{or} \quad (1 = 10I_1 - 6I_2)(1)$$
$$60 = -120I_1 + 180I_2 \qquad (1 = -2I_1 + 3I_2)(5)$$

step 4: Solve for I_2. Multiply the lower equation by 5 and add.

$$6 = 9I_2$$
$$I_2 = 0.667 \text{ A}$$

step 5: Solve for I_1. Substitute for I_2 in the equation

$$1 = -2I_1 + 3I_2$$
$$I_1 = 0.5 \text{ A}$$

A substitution of numbers in the other formula will check these results.

nodal analysis

A junction or point where two or more currents are entering or leaving is called a *node*. Kirchhoff's current law states that *the algebraic sum of currents entering a node is equal to zero*. The conventional methods of nodal analysis assumes that a positive current is always entering a node terminal and a negative current is leaving. For example, in Fig. 6–22, the node terminal is denoted by a heavy dot, and the current equation is

$$+I - I_1 - I_2 = 0$$

The other problem that exists in nodal analysis is the expression of nodal voltages. Refer to Fig. 6–23. Note that current I is flowing through R from right to left. The equation for the current through R using Ohm's law is

$$I = \frac{E_1 - E_2}{R}$$

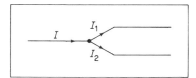

figure 6–22
currents at a node

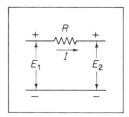

figure 6–23
node voltages

Note that, since I is positive in the direction shown, E_1 must be greater than E_2; otherwise, current I will reverse in the flow through R. An illustrative problem will demonstrate the theory.

---

sample problem

Determine E_a and E_b, using nodal analysis for the given circuit.

$I = 30$ ma	$E_2 = 72$ v	$R_2 = 15$ kΩ
$E_1 = 20$ v	$R_1 = 10$ kΩ	$R_3 = 24$ kΩ
		$R_4 = 96$ kΩ

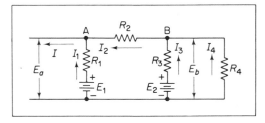

solution

step 1: Set up nodes A and B on the diagram. The current equations at these nodes are

Node A:
$$I = I_1 + I_2$$

Node B:
$$I_2 = I_3 + I_4$$

step 2:   Evaluate the currents in terms of the voltages.

$$I_1 = \frac{E_a - E_1}{R_1} \qquad I_2 = \frac{E_a - E_b}{R_2}$$

$$I_3 = \frac{E_b - E_2}{R_3} \qquad I_4 = \frac{E_b}{R_4}$$

step 3:   Write the two node equations in terms of voltages.

$$I = \frac{E_a - E_1}{R_1} + \frac{E_a - E_b}{R_2}$$

$$\frac{E_a - E_b}{R_2} = \frac{E_b - E_2}{R_3} + \frac{E_b}{R_4}$$

step 4:   Simplify these equations for simultaneous linear operation.

$$I + \frac{E_1}{R_1} = E_a \left(\frac{1}{R_1} + \frac{1}{R_2} \right) - E_b \left(\frac{1}{R_2} \right)$$

$$\frac{E_2}{R_3} = -E_a \left(\frac{1}{R_2} \right) + E_b \left(\frac{1}{R_2} + \frac{1}{R_3} + \frac{1}{R_4} \right)$$

step 5:   Substitute values for the symbols.

$$30 + 2 = E_a(\tfrac{1}{10} + \tfrac{1}{15}) - E_b(\tfrac{1}{15})$$

$$3 = -E_b(\tfrac{1}{15}) + E_a(\tfrac{1}{15} + \tfrac{1}{24} + \tfrac{1}{96})$$

step 6:   Simplify and solve for E_a and E_b.

$$32 = E_a(\tfrac{1}{6}) - E_b(\tfrac{1}{15})$$

$$3 = -E_a \left(\frac{1}{15} \right) + E_b \left(\frac{1}{8.42} \right)$$

Multiplying the top equation by 0.2 and the lower equation by 0.5 results in

$$6.4 = E_a(\tfrac{1}{30}) - E_b(\tfrac{1}{75})$$

$$1.5 = -E_a \left(\frac{1}{30} \right) + E_b \left(\frac{1}{16.84} \right)$$

$$E_b = \frac{7.9}{0.0461} = 171.5 \text{ v}$$

$$E_a = 6 \left(32 + \frac{171.5}{15} \right)$$

$$E_a = 260.58 \text{ v}$$

The reader should verify the solution, using the suitable network theorem.

problems

1. The arms of a π network are

$R_a = 425\ \Omega,$ $R_b = 575\ \Omega,$ $R_c = 1000\ \Omega$

Find the equivalent T network.

2. The arms of a π network are

$R_a = 650\ k\Omega,$ $R_b = 325\ k\Omega,$ $R_c = 975\ k\Omega$

Find the equivalent T network.

3. The arms of a T network are

$R_1 = 700\ \Omega,$ $R_2 = 2800\ \Omega,$ $R_3 = 3500\ \Omega$

Find the equivalent π network.

4. The arms of a T network are

$R_1 = 45\ k\Omega,$ $R_2 = 15\ k\Omega,$ $R_3 = 60\ k\Omega$

Find the equivalent π network.

5. Given the following network, find: (a) equivalent T, (b) equivalent π.

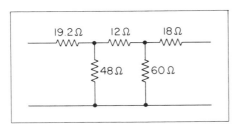

6. Given the following network, find: (a) equivalent T, (b) equivalent π.

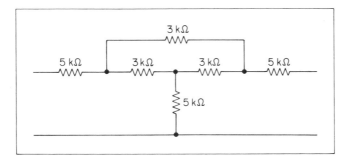

7. Given the circuit shown, determine the branch currents I_1 and I_2, using the superposition theorem. *Hint*: Convert inner π to a T.

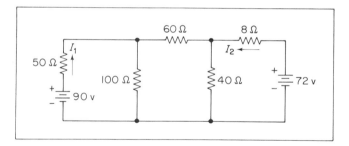

8. Given the circuit shown, determine the branch currents, using the super-position theorem.

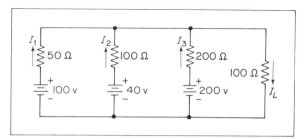

9 Using Thevenin's theorem, determine the current I_L through the load.

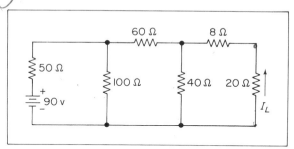

10. Using Thevenin's theorem, determine the current I_L through the load.

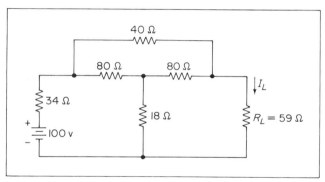

11. Find the current through the load, using (a) the superposition theorem; (b) Thevenin's theorem.

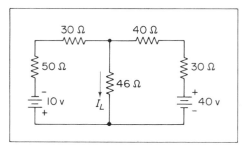

12. Using Norton's theorem, determine the load current I_L.

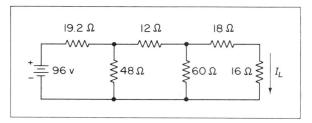

13. Solve for the output voltage, using Millman's theorem.

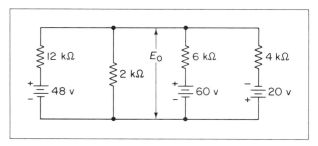

14. Solve for the output voltage, using Millman's theorem.

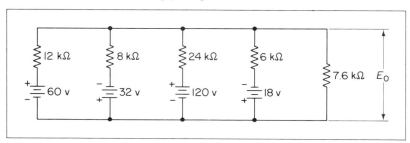

15. Determine the value of R_L for maximum power transfer to the load, and determine the load power.

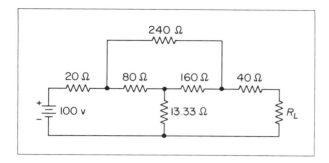

16. Determine the value of R_L for maximum power transfer, and evaluate the power.

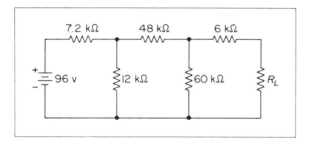

17. Determine the value of R_L for maximum power transfer, and evaluate the power.

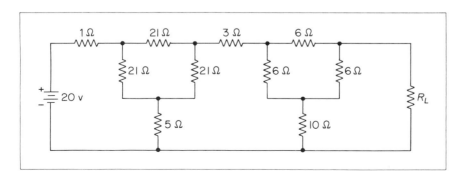

18. Determine I_1 and I_2 by loop circuit analysis.

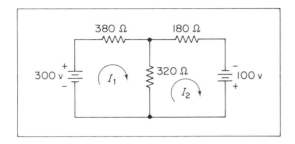

19. Write the loop equations for the circuit shown.

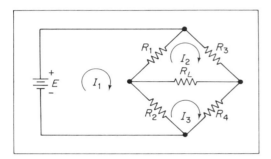

20. Determine the source currents I_1 and I_2 by loop circuit analysis.

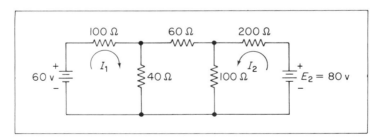

21. Determine E_a and E_b, using nodal analysis.

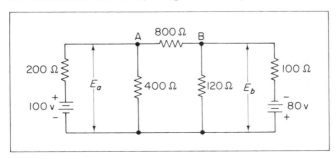

22. Determine E_a and E_b, using nodal analysis.

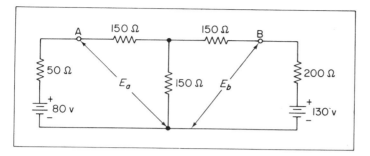

7

determinants

introduction

The analysis of circuits usually requires the solution of linear simultaneous equations. One method of solving linear equations is to substitute for one of the variables and thus eliminate this variable. This method of substitution and elimination is relatively simple but may be extremely laborious for a multiloop network. The simplified technique that is used to solve multimesh networks is called *determinants* and *matrix algebra*. It will be noted that although determinants and matrices bear some definite relationship to each other, the application of either technique to a network problem differs, and care must be used to avoid confusion between the two techniques.

Let us assume that we have the following system of two linear equations for which I_1 and I_2 are the unknowns. Thus,

$$R_1 I_1 + R_2 I_2 = E_1$$
$$R_3 I_1 + R_4 I_2 = E_2$$

In these two equations, usually the R parameters are known as well as the voltages E_1 and E_2. Using the method of substitution and elimination, multiply the top equation by R_4 and the bottom equation by R_2 and subtract the lower from the upper, obtaining

$$(R_1 R_4 - R_3 R_2) I_1 = E_1 R_4 - E_2 R_2$$

or

$$I_1 = \frac{E_1 R_4 - E_2 R_2}{R_1 R_4 - R_3 R_2}$$

The current I_1 could have been evaluated by the technique of determinants. Consider the two linear equations

$$E_1 = I_1 Z_{11} + I_2 Z_{12}$$

$$E_2 = I_1 Z_{21} + I_2 Z_{22}$$

Note that I_1 and I_2 are the variables, and the Z and E values are constants. To set up the determinant and to solve for I_1, the denominator is then a square array consisting of the coefficients of the variables. Thus,

$$\Delta = \begin{vmatrix} Z_{11} & Z_{12} \\ Z_{21} & Z_{22} \end{vmatrix} \quad \ldots \text{rows}$$

$$\vdots$$

columns

The numerator array is formed in a similar manner, except that if it is I_1 for which we wish to solve, the coefficients of I_1 are removed and replaced by the corresponding E terms. It should be noted that the E terms and the Z terms carry their own signs into the array. Thus,

$$\text{Numerator of } I_1 = \begin{vmatrix} E_1 & Z_{12} \\ E_2 & Z_{22} \end{vmatrix}$$

$$\text{Numerator of } I_2 = \begin{vmatrix} Z_{11} & E_1 \\ Z_{21} & E_2 \end{vmatrix}$$

Once the arrays have been established, the evaluation of the determinant is performed in the following manner. Draw diagonal arrows from E_1 to Z_{22} and from Z_{11} to Z_{22}. Thus,

$$I_1 = \frac{\begin{vmatrix} E_1 & Z_{12} \\ E_2 & Z_{22} \end{vmatrix}}{\begin{vmatrix} Z_{11} & Z_{12} \\ Z_{21} & Z_{22} \end{vmatrix}} = \frac{E_1 Z_{22} - Z_{12} E_2}{Z_{11} Z_{22} - Z_{12} Z_{21}}$$

The product of these terms along the diagonal is positive. Construct diagonals going upwards from E_2 and Z_{21} as shown. The product of these terms is also positive but subtracted from the first term. Thus, the solution to the determinant is shown above. In a similar manner, the value of I_2 can also be found. Thus,

$$I_2 = \frac{\begin{vmatrix} Z_{11} & E_1 \\ Z_{21} & E_2 \end{vmatrix}}{\begin{vmatrix} Z_{11} & Z_{12} \\ Z_{21} & Z_{22} \end{vmatrix}} = \frac{E_2 Z_{11} - E_1 Z_{21}}{Z_{11} Z_{22} - Z_{12} Z_{21}}$$

These particular determinants are referred to as *second-order* determinants, since each one contains two rows and two columns. It should be noted when one is using determinants that the columns of the determinant must be placed in the same order as indicated by the original equations.

examples

Evaluate the following determinants:

(a) $\begin{vmatrix} 2 & 3 \\ 5 & 7 \end{vmatrix}$ (b) $\begin{vmatrix} 4 & -2 \\ 0 & -3 \end{vmatrix}$ (c) $\begin{vmatrix} 5 & -2 \\ 3 & 7 \end{vmatrix}$

The use of determinants is not limited solely to solving two simultaneous linear equations. Determinants can be applied to any number of simultaneous linear equations.

Consider the following three linear simultaneous equations.

$$Z_{11}I_1 + Z_{12}I_2 + Z_{13}I_3 = E_1$$
$$Z_{21}I_1 + Z_{22}I_2 + Z_{23}I_3 = E_2$$
$$Z_{31}I_1 + Z_{32}I_2 + Z_{33}I_3 = E_3$$

The method of setting up the determinant configuration is similar to the procedure for the two simultaneous equations. The coefficients of I_1 form the first column of the denominator, the coefficients of I_2 form the second column, etc. The numerator is obtained, if the solution for I_1 is desired, by replacing the column containing the coefficients of I_1 by the elements to the right of the equals sign, or, in this case, the voltages. The determinant is

$$I_1 = \frac{\begin{vmatrix} E_1 & Z_{12} & Z_{13} \\ E_2 & Z_{22} & Z_{23} \\ E_3 & Z_{32} & Z_{33} \end{vmatrix}}{\begin{vmatrix} Z_{11} & Z_{12} & Z_{13} \\ Z_{21} & Z_{22} & Z_{23} \\ Z_{31} & Z_{32} & Z_{33} \end{vmatrix}}$$

A simplified method for the evaluation of a third-order determinant requires recopying the first two columns and placing them next to the third column of the determinant.

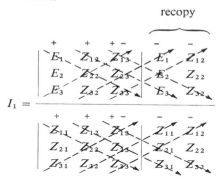

The solution of a third-order determinant requires six terms in both the numerator and denominator. The first three terms are found by drawing diagonal arrows from E_1, Z_{12}, and Z_{13} of the original determinant in the numerator from upper left to the lower right. Multiply the coefficients along each arrow and sum the resultants. Thus,

$$\Sigma_1 = E_1 Z_{22} Z_{33} + E_2 Z_{13} Z_{32} + E_3 Z_{12} Z_{23}$$

The last three terms are found by drawing diagonal dotted arrows from the lower left to the upper right. Multiply the coefficients along each dotted arrow and sum the resultants. Thus,

$$\Sigma_2 = E_3 Z_{22} Z_{13} + E_1 Z_{32} Z_{23} + E_2 Z_{33} Z_{12}$$

The value of the numerator when determinants are used is

$$(\Delta) \text{ numerator} = \Sigma_1 - \Sigma_2$$

or

$$\Delta_n = (E_1 Z_{22} Z_{33} + E_2 Z_{13} Z_{32} + E_3 Z_{12} Z_{23})$$
$$- (E_3 Z_{22} Z_{13} + E_1 Z_{32} Z_{23} + E_2 Z_{33} Z_{12})$$

The value of the denominator when determinants are used is

$$\Delta_d = (Z_{11} Z_{22} Z_{33} + Z_{12} Z_{23} Z_{31} + Z_{13} Z_{21} Z_{32})$$
$$- (Z_{31} Z_{22} Z_{13} + Z_{32} Z_{23} Z_{11} + Z_{33} Z_{21} Z_{12})$$

The value of the numerator for I_2 when determinants are used is

$$\Delta = (Z_{11} E_2 Z_{33} + E_1 Z_{23} Z_{31} + Z_{13} Z_{21} E_3)$$
$$- (Z_{31} E_2 Z_{13} + E_3 Z_{23} Z_{11} + Z_{33} Z_{21} E_1)$$

The value of the numerator for I_3 when determinants are used is

$$\Delta = (Z_{11}Z_{22}E_3 + Z_{12}Z_{31}E_2 + E_1Z_{21}Z_{32})$$
$$- (Z_{31}Z_{22}E_1 + Z_{32}E_2Z_{11} + E_3Z_{21}Z_{12})$$

The denominator determinant is common to the solutions for I_1, I_2, and I_3.

---

sample problem

$$\Delta = \begin{vmatrix} 0 & 8 & 4 \\ 2 & 3 & 5 \\ 1 & 2 & -2 \end{vmatrix}$$

---

solution

$$\Delta = \begin{vmatrix} 0 & 8 & 4 \\ 2 & 3 & 5 \\ 1 & 2 & -2 \end{vmatrix} \begin{matrix} 0 & 8 \\ 2 & 3 \\ 1 & 2 \end{matrix}$$

$\Sigma_1 = 0 + 40 + 16 = 56$

$\Sigma_2 = 12 + 0 + (-32) = -20$

$\Delta = 56 - (-20) = 76$

---

sample problem

Given the following three equations:

$$4I_1 - 2I_2 + 6I_3 = 10$$
$$-2I_1 + 7I_2 - 3I_3 = 0$$
$$6I_1 - 3I_3 + 10I_3 = 0$$

find I_1.

---

solution

$$\Delta_d = \begin{vmatrix} 4 & -2 & 6 \\ -2 & 7 & -3 \\ 6 & -3 & 10 \end{vmatrix} \begin{matrix} 4 & -2 \\ -2 & 7 \\ 6 & -3 \end{matrix}$$

$= (280 + 36 + 36) - (252 + 36 + 40)$

$= 352 - 328 = 24$

$$\Delta_n = \begin{vmatrix} 10 & -2 & 6 \\ 0 & 7 & -3 \\ 0 & -3 & 10 \end{vmatrix} \begin{matrix} 10 & -2 \\ 0 & 7 \\ 0 & -3 \end{matrix}$$

$$= (700+0+0)-(0+90+0)$$

$$= 610$$

determinant of order n

A determinant may be reduced in order by the method of minors. It should be noted that the number of terms in the solution of any determinant is defined as equal to the order of the determinant factorial. Mathematically, the relationship is

$$\text{No. of terms} = n!$$

From the foregoing discussion, a fifth-order determinant will have 120 terms in the resultant solution.

Consider a third-order determinant as shown.

$$\Delta = \begin{vmatrix} a_1 & a_2 & a_3 \\ b_1 & b_2 & b_3 \\ c_1 & c_2 & c_3 \end{vmatrix}$$

The minor of a_1, designated by Δ_1, is

$$\Delta_1 = \begin{vmatrix} b_2 & b_3 \\ c_2 & c_3 \end{vmatrix}$$

and is found from the original determinant by removing the row and column of a given element. Thus,

$$\Delta = \begin{vmatrix} -a_1-a_2-a_3- \\ b_1 \quad b_2 \quad b_3 \\ c_1 \quad c_2 \quad c_3 \end{vmatrix}$$

The minor of a_2, designated by Δ_2, is

$$\Delta_2 = \begin{vmatrix} b_1 & b_3 \\ c_1 & c_3 \end{vmatrix}$$

This reduction technique can be applied to any order determinant. The following mathematical relationship should be applied to take care of any sign convention.

$$\Delta = \Sigma \, (-1)^{i+j} |M_{ij}| a_{ij}$$

where

$|M_{ij}|$ = minor of the element under investigation.
a_{ij} = element of interest.

Using this relationship, we find that the solution of Δ is

$$\Delta = a_1 \begin{vmatrix} b_2 & b_3 \\ c_2 & c_3 \end{vmatrix} - a_2 \begin{vmatrix} b_1 & b_3 \\ c_1 & c_3 \end{vmatrix} + a_3 \begin{vmatrix} b_1 & b_2 \\ c_1 & c_2 \end{vmatrix}$$

$$\Delta = a_1(b_2 c_3 - b_3 c_2) - a_2(b_1 c_3 - b_3 c_1) + a_3(b_1 c_2 - b_2 c_1)$$

---

sample problem

Given

$$\Delta = \begin{vmatrix} 3 & 5 & 2 & 7 \\ 0 & 4 & 3 & 1 \\ 9 & 0 & 2 & 5 \\ 8 & 2 & 0 & 3 \end{vmatrix}$$

---

solution

step 1:   Apply the technique of minors. Thus, the first minor is

$$\Delta_1 = 3 \begin{vmatrix} 4 & 3 & 1 \\ 0 & 2 & 5 \\ 2 & 0 & 3 \end{vmatrix} = 150$$

step 2:   The total second minor is

$$\Delta_2 = -5 \begin{vmatrix} 0 & 3 & 1 \\ 9 & 2 & 5 \\ 8 & 0 & 3 \end{vmatrix} = -115$$

step 3:   The third minor is

$$\Delta_3 = 2 \begin{vmatrix} 0 & 4 & 1 \\ 9 & 0 & 5 \\ 8 & 2 & 3 \end{vmatrix} = 140$$

step 4: The fourth minor is

$$\Delta_4 = -7 \begin{vmatrix} 0 & 4 & 3 \\ 9 & 0 & 2 \\ 8 & 2 & 0 \end{vmatrix} = -826$$

step 5: The total solution is

$$\Delta = \Delta_1 + \Delta_2 + \Delta_3 + \Delta_4$$

$$\Delta = -651$$

problems

1. Evaluate the following determinants.

(a) $\begin{vmatrix} 4 & 2 \\ 2 & 5 \end{vmatrix}$ (b) $\begin{vmatrix} 3 & -4 \\ -4 & 6 \end{vmatrix}$

(c) $\begin{vmatrix} 3 & 0 \\ -2 & 5 \end{vmatrix}$ (d) $\begin{vmatrix} 5 & -7 \\ -2 & 4 \end{vmatrix}$

2. Evaluate the following determinants.

(a) $\begin{vmatrix} 5 & 2 & 11 \\ 2 & 4 & 1 \\ 11 & 1 & 7 \end{vmatrix}$ (b) $\begin{vmatrix} 2 & 1 & 0 \\ 1 & 5 & -3 \\ 0 & -3 & 4 \end{vmatrix}$

3. Solve for x and y, using determinants.

(a) $2x + 3y = 8$
    $x - 2y = 6$

(b) $3x - 5y = 1$
    $4x + 7y = 3$

(c) $3x - 5y = 0$
    $5y - 7 = 9x$

(d) $4x - 7y = 3$
    $x + 2y = 5$

4. Solve for the unknowns, using determinants.

(a) $2x + 5y \quad = 5$
    $x + 8y - 3z = 9$
    $-3y + 5z = -4$

(b) $12x - 3y + 4z = 10$
    $-3x + 3y - 5z = 4$
    $4x - 5y + 2z = -7$

5. Solve for e_1, e_2, and e_3, using determinants.

$$2e_1 - e_2 + 3e_3 = 7$$
$$3e_1 + 2e_2 - e_3 = 5$$
$$e_1 + 3e_2 + 2e_3 = 10$$

6. Solve for the unknowns, using determinants.

(a) $3i_1 - i_2 + i_3 = 5$
$-i_1 + 5i_2 - 3i_3 = 12$
$i_1 - 3i_2 + 2i_3 = 8$

(b) $3i_1 \quad - i_3 + i_4 = 5$
$4i_2 - i_3 + 5i_4 = 6$
$-i_1 - i_2 + 6i_3 - 3i_4 = 0$
$0 + 5i_2 - 3i_3 + 2i_4 = 3$

7. Evaluate the unknown currents by determinants.

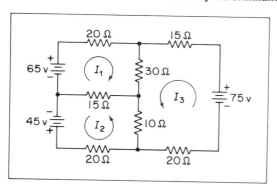

8. Evaluate the unknown currents, using determinants.

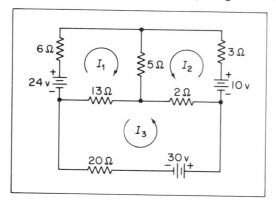

9. Evaluate the unknown voltages E_a and E_b by determinants.

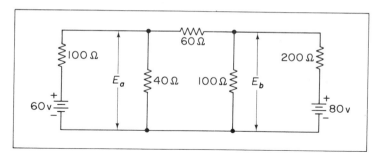

10. Evaluate the unknown voltages by means of determinants.

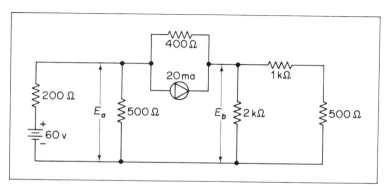

8

capacitance

introduction

The connection of a battery source across a network of resistances resulted in a current flow. Note that resistance is defined as that property of a circuit element that opposes the current flow.

There are electrical devices that are linear and passive but operate on the basis of *changes in voltage or current*. Under ideal conditions, these devices do not dissipate energy as a resistor does, but store energy until required to return it. In this chapter, we shall consider one of these devices, called a *capacitor*. Capacitance is the property of an electrical circuit element that opposes *the flow of current and any change in voltage across it*.

electric field

Consider the electric field that exists around a charged sphere, as shown in Fig. 8–1. The electric field directed radially outward is represented by the lines of force called flux lines. The definition of the electric field intensity at a point is

$$\mathcal{E} = \frac{F \text{ (newtons)}}{Q \text{ (coulombs)}} \tag{8-1}$$

A force F_t directed r_t meters away from Q to a unit test charge Q_t is determined by Coulomb's law to be equal to

$$F_t = \frac{QQ_t}{r_t^2} = \frac{Q}{r_t^2} \tag{8-2}$$

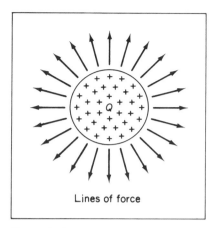

figure 8–1
electric field

flux density

It is evident that the stronger the electric field, the denser will be the number of lines of force or flux. Consider the electric field intensity existing between two parallel plates, as shown in Fig. 8–2. Note that the flux lines always are directed from a positively to a negatively charged body. The symbol for electric flux is the Greek letter ψ (psi), and the symbol for the flux per unit area is represented by the letter D. Since the flux is uniform between two parallel plates, the equation for the density of the electric field becomes

$$D = \frac{\psi}{A} \frac{\text{newtons}}{\text{unit area}} \tag{8–3}$$

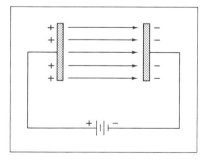

figure 8–2
electric field between two parallel plates

Since the electric field is uniform between the parallel plates, a constant force exists on a charged particle in this region. The work performed in moving a charged particle from one plate to the other is

$$W = Fd \tag{8–4}$$

The magnitude of the volt was previously defined as work per unit charge, or

$$E = \frac{W}{Q} \text{ volts} \tag{8–5}$$

Then, substituting both for force and work results in

$$\mathscr{E} = \frac{E}{d} \tag{8–6}$$

where

$\mathscr{E}$ = electric field strength in newtons/coulomb and also in volts/meter.
E = potential difference in volts.
d = distance between the plates in meters.

capacitance

Consider two identical parallel plates without charge separated by a distance d as shown in Fig. 8–3. The insulating material that exists between these two plates is air. When the switch is closed, negative charges accumulate on the plate attached to the negative electrode and positive charges accumulate

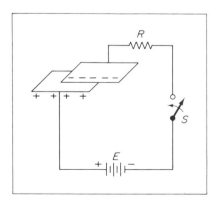

figure 8–3
parallel plate capacitor

on the other plate. An electric field is established between the two plates with an eventual net current flow of zero.

If E can be increased, a value will be reached whereby the insulating material or dielectric is ruptured between the two plates. There is conduction as if the two plates were directly shorted. Some dielectrics can withstand greater potential differences than others before rupturing.

The values of dielectric strength listed in Table 8–1 are typical average values for some of the common dielectric materials.

table 8–1   relative permittivity of dielectric constants

material	dielectric constant
vacuum	1.0000
air	1.0006
bakelite	7.000
glass	7.500
mica	8.000
oil (transformer)	3.000
paraffined paper	2.500
porcelain	6.000
rubber	3.000
teflon	2.000

After the source voltage applied to the two parallel plates has been removed, one must short the two plates together externally to permit neutralization of the charges; otherwise, a residual electrostatic field charge will remain between the two plates.

A circuit component that consists of two parallel plates separated by a dielectric is called a capacitor. Capacitance is the ability of a capacitor to store charge between its plates. *A capacitor has a capacitance of 1 farad*[1] *when a charge of 1 coulomb is placed between the two plates by a potential difference of 1 volt.* Mathematically, the relationship is expressed by

$$C = \frac{Q \text{ (coulombs)}}{E \text{ (volts)}} \text{ farads} \qquad (8\text{–}7)$$

where

C = capacitance in farads.
Q = charge in coulombs.
E = potential difference in volts.

[1]Named in honor of Michael Faraday.

Note, therefore, that the measure of capacity is its ability to store charge. The larger the charge stored by the dielectric for a given source voltage, the greater the capacitance. If the area of the plates is increased, the electric field intensity will increase accordingly, resulting in a greater capacitance. Thus, the capacitance will vary directly with the area of the plates. Consider the distance between the plates to be variable. Decreasing the distance between the plates increases the strength of the electric field and consequently increases the storage ability of the capacitor. Thus, the capacitance will vary inversely with the distance between the two plates. It is evident from the foregoing discussion that the capacitance of parallel plates is directly proportional to the dielectric and to the area of the plates and inversely proportional to the distance between the plates. Mathematically this relationship can be expressed by

$$C = \epsilon \frac{A}{d} \text{ farads} \tag{8-8}$$

where

C = capacitance in farads.
ϵ = permittivity of the dielectric in MKS units.
A = area of each plate in square meters.
d = distance between plates in meters.

The absolute permittivity of free space is equal to 8.85×10^{-12} in MKS units. The relative permittivity or dielectric constant is defined as the ratio of the absolute permittivity of the material with respect to the absolute permittivity of free space. The symbol κ (kappa) represents the relative permittivity. Thus, the relationship for the capacitance of two parallel plates using any dielectric is

$$C = \frac{8.85 \times 10^{-12} \kappa A}{d} \text{ farads} \tag{8-9}$$

where

κ = dielectric constant.
A = area of the plate in square meters.
d = distance between plates in meters.

Using the conversion methods to convert square meters to square inches and meters to inches results in

$$C = \frac{2.25 \times 10^{-7} \kappa A}{d} \text{ farads} \tag{8-10}$$

An illustrative problem will demonstrate the theory.

sample problem

The parallel plates of a capacitor are 2 mm apart, and each plate has an area of 0.04 square meter. The dielectric between the plates is mica. The source voltage is 400 volts. Find: (a) the capacitance, (b) the electric field strength, (c) the charge stored.

solution

step 1: From Table 8–1, the dielectric constant is 5. Calculate C.

$$C = \frac{8.85 \times 10^{-12} \kappa A}{d}$$

$$C = \frac{8.85 \times 10^{-12}(5)(0.04)}{2 \times 10^{-3}}$$

$$C = 885 \text{ pF}$$

step 2: Calculate the field strength.

$$\mathscr{E} = \frac{E}{d}$$

$$\mathscr{E} = \frac{400}{2 \times 10^{-3}}$$

$$\mathscr{E} = 200 \times 10^{3} \text{ volts/meter}$$

step 3: Calculate Q.

$$Q = CE$$

$$Q = 8.85 \times 10^{-10} \times 400$$

$$Q = 0.354 \ \mu C$$

capacitors in series and parallel

The symbol used for a capacitor is shown in Fig. 8–4. Connect three capacitors in series as shown in Fig. 8–5. When the switch S is closed, the electrons align themselves on the negative terminals, as shown in Fig. 8–5. The number of electrons moved through each capacitor is the same, making the three charges Q_1, Q_2, and Q_3 all equal and symbolized by Q.

The application of Kirchhoff's voltage law yields

$$E = E_1 + E_2 + E_3$$

figure 8–4
symbol for capacitor

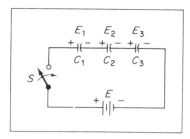

figure 8–5
capacitors in series

Since $E = Q/C$, then substituting yields

$$\frac{Q}{C} = \frac{Q}{C_1} + \frac{Q}{C_2} + \frac{Q}{C_3}$$

The equivalent capacitance of the network is equal to

$$\frac{1}{C} = \frac{1}{C_1} + \frac{1}{C_2} + \frac{1}{C_3}$$

The equation may be generalized for any number of capacitors in series. Thus,

$$\frac{1}{C} = \frac{1}{C_1} + \frac{1}{C_2} + \frac{1}{C_3} + \cdots + \frac{1}{C_n} \qquad (8\text{–}11)$$

Note, therefore, that *capacitors in series* are evaluated in the same manner as *resistors in parallel*.

sample problem

Determine the total capacitance of the network shown and the voltage drop across each capacitor.

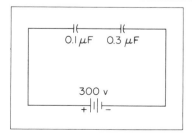

solution

step 1: Calculate the total capacitance C_T.

$$\frac{1}{C_T} = \frac{1}{C_1} + \frac{1}{C_2}$$

$$C_T = \frac{C_1 C_2}{C_1 + C_2} = \frac{0.1 \times 0.3}{0.4} \ \mu F$$

$$C_T = 0.075 \ \mu F$$

step 2: Calculate Q.

$$Q = E C_T = 300 \times 0.075 \times 10^{-6}$$

$$Q = 22.5 \ \mu C$$

step 3: Calculate E_1 and E_2.

$$E_1 = \frac{Q}{C_1} = \frac{22.5 \times 10^{-6}}{0.1 \times 10^{-6}}$$

$$E_1 = 225 \ v$$

$$E_2 = 300 - 225 = 75 \ v$$

capacitors in parallel

When capacitances are connected in parallel, the end result is that the areas of the plates have increased with a consequent increase in capacitance. Refer to Fig. 8–6. Note that the total charge Q_T is equal to the sum of the individual charges. Thus,

$$Q_T = Q_1 + Q_2 + Q_3$$

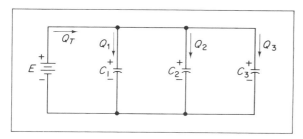

figure 8–6
capacitors in parallel

It is evident that the same voltage exists across each capacitor. Therefore,

$$Q_T = EC_1 + EC_2 + EC_3 = EC_T$$

Thus the total capacitance is the sum of the individual capacitors. Consequently,

$$C_T = C_1 + C_2 + C_3$$

A 100-pF capacitor in parallel with a 50-pF capacitor, for example, yields a total capacitance of 150 pF. It is obvious that *capacitors connected in parallel* are treated in the same manner as *resistors in series*.

The equation may be generalized for any number of capacitors in parallel as

$$C_T = C_1 + C_2 + C_3 + \cdots + C_n \qquad (8\text{--}12)$$

types of capacitors

Capacitors may be classified according to voltage ratings, dielectric, frequency, etc. Commercial capacitors are commonly referred to by the dielectric material. Typical dielectrics used are

1. Mica	2. Paper	3. Ceramic
4. Electrolytic	5. Variable	6. Integrated circuit
7. Air	8. Oil filled	

mica capacitors

Sheets of tinfoil separated by mica form mica capacitors. Alternate strips of tinfoil are connected together to form one plate of the capacitor, as shown in Fig. 8–7. The internal elements of the capacitors are enclosed in a molded Bakelite or waterproof plastic case. Mica capacitors are quite commonly used in radio and television applications because of their ability to withstand relatively high voltages and maintain excellent frequency characteristics. The usual capacitance range of capacitors of this manufacture is from 25 to 500 pF.

paper capacitors

In this construction, long narrow sheets of alternate layers of aluminum foil and wax-impregnated paper are rolled into a compact layer or cylinder, as shown in Fig. 8–8.

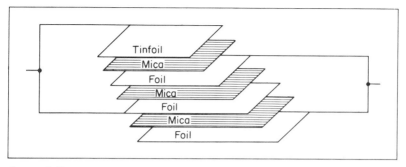

figure 8–7
mica capacitors

The internal elements are enclosed in a cardboard container impregnated with wax or in a waterproof case. The aluminum sheet is connected to one side of the capacitor, usually the ground or reference point. This side usually has a black band at that end of the capacitor. It is usually advisable to connect this side to ground or low potential, but there is no required polarity to the capacitor.

ceramic capacitors

The construction of this type of capacitor utilizes silver or copper coatings on each side of a ceramic base. An insulating coating is then applied over the base and silver coating. This type of capacitor is used in miniaturized electronic components.

electrolytic capacitors

The construction of this type of capacitor utilizes aluminum foil as one plate and an alkaline electrolyte as the other plate. The application of a dc source voltage during manufacture causes the electrolytic action to produce a thin layer of aluminum oxide, which acts as a dielectric. The layer is extremely thin, leading to high values of capacitance.

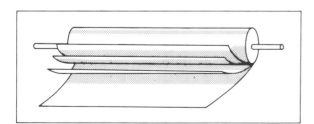

figure 8–8
paper capacitor

Electrolytic capacitors must be connected to a dc source according to the polarity shown on the capacitor. Reversal of the polarity connection will damage and burn out the capacitor. The advantages of this type of capacitor are high capacitance values and relatively small size. The disadvantages are the required polarity and high leakage of current.

variable capacitors

Variable capacitors are constructed by placing movable plates into fixed plates separated by air. The fixed plates are connected together and form the stator. The movable plates are connected to a shaft and form the rotor. The capacitance is varied by rotating the rotor to mesh with the stator. When the plates are fully meshed, the capacitance is maximum; out of mesh gives minimum capacitance. For extremely high-voltage operation, the stator and rotor may be placed in a vacuum.

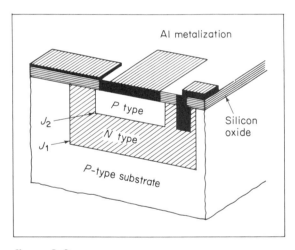

figure 8–9
diffused junction capacitor

integrated circuit capacitors

Integrated circuit capacitors may readily be formed by using the junction capacitance of a reversed P-N junction. A cross-sectional view of a junction capacitance is shown in Fig. 8–9.

Whenever a semiconductor junction is formed, a junction capacitance results. The capacitor is formed by reverse-biasing the P-type and N-type regions, labelled junction J_2 on Fig. 8–9. An additional capacitor is formed by junction J_1. The capacitance value of such a structure is given by

$$C = 0.885 \frac{KA}{d} \text{ pF}$$

where

A = junction area in square centimeters.
K = dielectric constant of insulating material.
d = depletion layer thickness in centimeters.

The dielectric constant K for silicon is 12. Note that capacitance C is a direct function of the junction area and inversely proportional to the thickness of the depletion region. The voltage input must have the negative terminal applied to the substrate, since the junction J_2 must be reverse-biased for proper operation.

RC charging circuit

Consider the series RC circuit shown in Fig. 8–10. At the instant the switch S is closed, the capacitor is a short circuit, since it takes a finite length of time for the charges to align themselves. The instantaneous or initial current in the circuit is then equal to

$$(\text{initial}) \ i = \frac{E}{R}$$

This current flow then is used to charge the capacitor, and a potential difference begins to form across the capacitor. As the voltage across the capacitor increases, the flow of current decreases. This sequence continues until the capacitor is fully charged to the applied voltage and the current flow in the circuit becomes equal to zero.

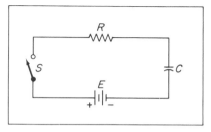

figure 8–10
RC series charge circuit

Consider the voltage across C to be equal to e_c and to vary directly as a function of time. Then the Kirchhoff's law equation for the circuit is

$$E = e_c + iR$$

After some time has elapsed, the capacitor charges to the applied voltage E, and the current becomes zero. Since the current varies with time, a graph of current versus time is shown in Fig. 8–11. The voltage e_c across C also is varying with time, as shown in Fig. 8–12.

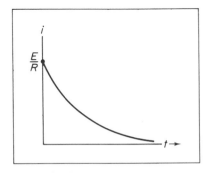

figure 8–11
charge current

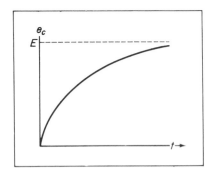

figure 8–12
e_c vs. t

analysis

The mathematical equations required to construct these graphs can be derived. The lowercase letters will denote instantaneous values. Thus,

$$q = Ce_c$$

Since C is a constant, the change in e_c must occur because of the change in q. Since both charge and capacitor voltage vary with time, the time rate change of charge is equal to

$$\frac{dq}{dt} = C\frac{de_c}{dt}$$

where

$\dfrac{dq}{dt}$ = time rate change of charge.

$\dfrac{de_c}{dt}$ = time rate change of capacitor voltage.

Again, current is defined as the time rate change of charge; then

$$i = \frac{dq}{dt}$$

Substituting this relationship yields

$$i = C\frac{de_c}{dt} \text{ amperes} \qquad (8\text{--}13)$$

This equation is extremely important because it interrelates voltage and current with the capacitance. Mathematically, the solution for e_c requires the integral of both sides of the equation. Thus,

$$e_c = \frac{1}{C}\int i\,dt$$

where the symbol $\int$ denotes the total sum of the varying currents with respect to time. Note, therefore, that the instantaneous charge on the capacitor is defined by

$$q = \int i\,dt$$

Consequently, the Kirchhoff's equation for the circuit becomes

$$E = iR + \frac{1}{C}\int i\,dt$$

This equation can be solved for the current i by using calculus. The solution is

$$i = \frac{E}{R}\epsilon^{-t/RC} \qquad (8\text{--}14)$$

The voltage drop across the resistor is

$$e_R = iR = E\epsilon^{-t/RC} \qquad (8\text{--}15)$$

The voltage across the capacitor is equal to

$$e_c = E - e_R = E(1 - \epsilon^{-t/RC}) \qquad (8\text{--}16)$$

time constant

The power to which ϵ is raised is t/RC. For the above equation to be valid, $R \times C$ must be equal to time. Dimensionally, R is in units of volts per ampere and C is in coulombs per volt. The dimension of a coulomb is ampere-second. Putting all of these together yields

$$RC = \left(\frac{E}{I} \right) \left(\frac{Q}{E} \right) = \frac{IT}{I}$$

$$RC = \text{time } (T)$$

The capacitor C does not affect the initial current, but appears only in the exponential term. The exponent determines the speed with which the current decays to zero. Thus, the effect of a varying C is shown in Fig. 8–13.

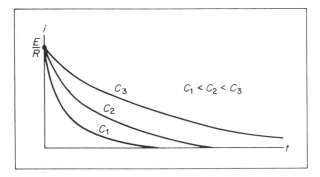

figure 8–13
effect of varying C

The product of RC is called the time constant. Thus

$$RC = \tau \qquad (8\text{–}17)$$

where

R is in ohms.
C is in farads.

The various exponential equations for a series RC circuit can be simplified as follows by using the time constant.

$$i = \frac{E}{R} \, \epsilon^{-t/\tau}$$

and

$$e_R = E\epsilon^{-t/\tau}$$

$$e_C = E(1 - \epsilon^{-t/\tau})$$

It is evident that there is a dependent relationship between time constant and the actual time. A graph called a universal curve can be plotted with the abscissa calibrated in terms of time constants rather than seconds. The ordinate is calibrated in terms of percentage of the applied voltage E. The curves are shown in Fig. 8–14. When the value of τ is exactly equal to t, then the current drops to 36.8 percent of its initial value, whereas the voltage across the capacitor rises to 63.2 percent of its final steady-state value.

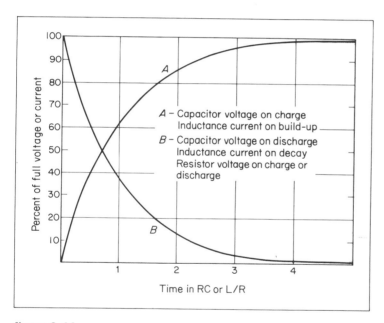

figure 8–14
RC time constant

sample problem

In the series *RC* circuit shown, what is the voltage across *C* 4 seconds after the switch is closed? Determine the time it takes the voltage across the capacitance to rise to 40 volts.

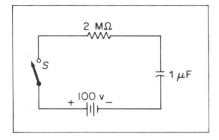

solution

step 1: Calculate τ.

$$RC = \tau = 2 \times 1 = 2 \text{ sec}$$

Note that 4 seconds corresponds to 2τ.

step 2: From the diagram of Fig. 8–14,

$$e_c = 86\% \text{ of } E = 86 \text{ v}$$

step 3: Determine t. From Fig. 8–14, $40 = 4\%$ of E.

$$t = 0.5\tau = 1 \text{ sec}$$

The same problem can be solved analytically. Thus,

$$e_c = E(1 - \epsilon^{-t/RC})$$
$$e_c = 100(1 - \epsilon^{-2}) = 100(1 - 0.135)$$
$$e_c = 86.5 \text{ v}$$

The time problem can be solved also. Thus,

$$40 = 100\,(1 - \epsilon^{-0.5t})$$
$$t = 1.02 \text{ sec}$$

RC discharge circuit

Consider the circuit shown in Fig. 8–15. Assume that switch S has been in position 1 for a length of time equal to or greater than five time constants. Capacitor C is fully charged to the applied voltage E, and the current flow in the circuit is zero. Move the switch to position 2. The charge existing across C appears as if there were a voltage in series with capacitor C. Note that the voltage across C must equal the Ri drop. Thus,

$$\int \frac{i\,dt}{C} = iR$$

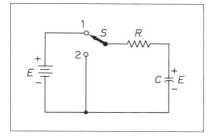

figure 8–15
discharging *RC* circuit

Solving this equation by calculus, we find that the instantaneous current *i* is

$$i = -\frac{E}{R}\epsilon^{-t/RC}$$ (8–18)

The voltage drop across the resistor at any instant of time is

$$e_R = -E\epsilon^{-t/RC} = e_c$$ (8–19)

A graph of these two equations is shown in Fig. 8–16.

The negative sign is a result of the current's flowing through the resistor in a direction opposite to the charge current. The time constant is obviously equal to the same quantity as the charge circuit unit. An illustrative problem will demonstrate the theory.

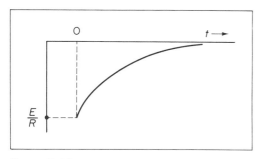

figure 8–16
discharge current vs. time

sample problem

In the circuit shown, what is the voltage across R at 0.16 sec after the switch is in position 2?

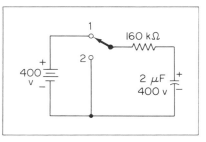

solution

step 1:   Calculate τ.

$$\tau = RC = 2 \times 10^{-6} \times 160 \times 10^{3}$$
$$\tau = 320 \text{ ms}$$

step 2:   Calculate e_R.

$$e_R = E\epsilon^{-0.16/0.32}$$
$$e_R = 400\epsilon^{-0.5}$$

From graph:
$$e_R = 60\% \text{ of } 400 \text{ v} = 240 \text{ v}$$

Analytically:
$$e_R = 0.6065(400) = 242.6 \text{ v}$$

sample problem

In the circuit shown, the switch S is in position 1. What is the potential difference across C when the switch is placed to position 2 after a time of 4 milliseconds?

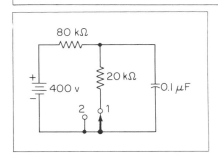

solution

step 1:   With S in position 1, the circuit can be Thevenized to determine the charge resistor and the voltage to which the capacitor can charge. Thus,

$$R_{th} = \frac{80 \times 20}{80 + 20} \, 10^3 = 16 \text{ k}\Omega$$

$$E_{oc} = \frac{400}{80 + 20} \, 20 = 80 \text{ v}$$

The resultant circuit is

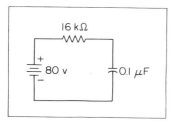

step 2:   The maximum voltage to which capacitor C can charge is 80 volts. If S is put to position 2, then the charging voltage is $(400 - 80)$. Thus,

$$e_c = 320(1 - \epsilon^{-t/RC})$$

When S is in position 2, the 20-kΩ resistor may be neglected, and the only resistor involved in the time constant is the 80-kilohm resistor.

$$RC = 80 \times 10^3 \times 0.1 \times 10^{-6} = 8 \text{ ms}$$

$$\frac{t}{RC} = 0.5$$

From graph:

$$\epsilon^{-0.5} = 0.6$$

Using the graph, we obtain

$$e_c = 320(1 - 0.6) = 128 \text{ v}$$

The total

$$e_c = 128 + 80 = 208 \text{ v}$$

energy stored in a capacitor

During the capacitor charge, current must flow. Energy is taken from the source. Some of this energy is dissipated in the form of heat across the series current-limiting resistor. The remainder of the energy is stored in the electric field of the capacitor.

Removal of a charged capacitor from the source of energy maintains the charge on the plates for a considerable length of time if there is negligible leakage between the plates. Shorting out the plates discharges the energy stored within the capacitor. A graph can be made on the rate at which a capacitor stores energy; one is shown in Fig. 8–17.

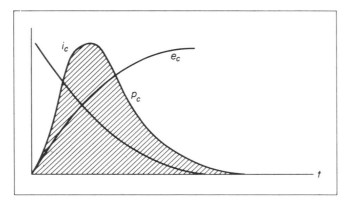

figure 8–17
power vs. time

The area covered by the p_c curve is dimensionally equal to watts multiplied by time, yielding joules. The total energy stored by the capacitor can also be found analytically by

$$w = \int_0^t p_c \, dt \text{ joules}$$

and

$$p_c = i_c e_c = C \frac{de_c}{dt} (e_c)$$

Then

$$w = \int_0^t Ce_c \frac{de_c}{dt} (dt) = \frac{1}{2} Ce_c^2$$

$$\boxed{w = \tfrac{1}{2} Ce_c^2 \text{ joules}} \tag{8–20}$$

problems

1. Find the strength of an electric field at a point 3.5 meters from a charge of 49 uC.

2. The electric field strength is 60 newtons/coulomb at a point x meters from a charge of 8 coulombs. Find the distance x.

3. Find the capacitance of a parallel plate capacitor if 2000 μC of charge is stored when 40 volts is applied.

4. Determine the charge of a 0.05-μF capacitor when 125 volts is applied.

5. Find the electric field strength between the parallel plates of a capacitor when a 500-mv potential is applied and the distance between the plates is 4 mm.

6. A 10-μF capacitor has 240 μC of charge stored between the plates. If the plates are 6 mm apart, find the electric field strength.

7. Find the capacitance of a parallel plate capacitor if the area of each plate is 0.06 m² and the distance between the plates is 3 mm. The dielectric is air.

8. Repeat Problem 7 with the dielectric of glass.

9. Determine the distance between the plates of a 0.04 μF capacitor. The area of the plates is 0.084 m². The dielectric is rubber.

10. The parallel plates of a capacitor are 0.6 mm apart and have an area of 0.036 m². The dielectric is glass. The applied voltage is 300 volts. Find: (a) the capacitance, (b) the electric field strength, (c) the charge.

11. For the circuit shown, find: (a) C_T; (b) the voltage across each capacitor.

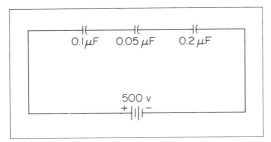

12. For the circuit shown, find: (a) C_T; (b) the voltage across each capacitor.

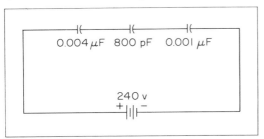

13. For the circuit shown, find: (a) C_T; (b) Q_1 and Q_2.

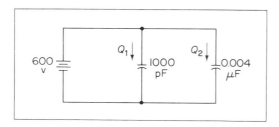

14. For the circuit shown with a value of Q_1 equal to 2 μC, find: (a) C_T; (b) Q_2 and Q_3.

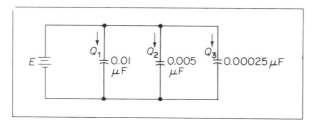

15. In the given circuit, find the voltage across C at each of the following times, assuming that switch S is just closed: (a) 4 ms, (b) 8 ms, (c) 20 ms.

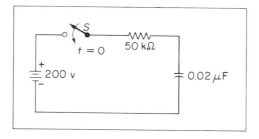

16. A 10-μF capacitor is to be charged through a 400-kilohm resistor from a 200-volt source. (a) What is the initial current? (b) What is the initial voltage across C? (c) What is the current 2 seconds after the switch is closed? (d) How long will it take the voltage across the capacitor to rise from 40 to 160 volts?

17. Given the following circuit with switch in position 1, find: (a) the voltage across C after 40 μsec; (b) the current in the circuit after 75 μsec. Capacitor C is fully charged. Switch S is in position 2. Find: (c) the voltage across R after 60 μsec; (d) the current in the circuit after 50 μsec.

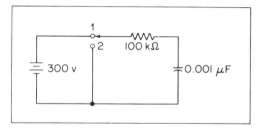

18. Given the following circuit, if the switch is closed, determine (a) the voltage across C after 0.5 ms; (b) the current in the circuit after 0.75 ms.

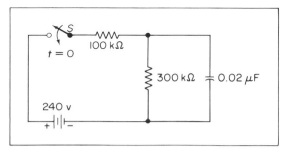

19. Given the following circuit, if the switch is in position 2, determine (a) the voltage across C after 0.5 ms, (b) the current in the circuit after 1 ms. With the switch in position 1, if capacitor C is fully charged, determine (c) the voltage across C after 0.5 ms, (d) the current in the circuit after 1 ms.

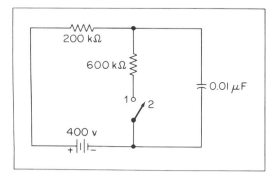

9
magnetic circuits

introduction

Electric motors, modern communication systems, radio and television etc. all employ the theory of magnetism in their operation.

A magnet is an object that can attract magnetic substances. The region around the magnet is known as the magnetic field. At any point in the space around the magnet, there is a state of stress that exerts a force on any magnetic pole brought into the region of the magnet. The direction that such a force takes indicates the direction of the magnetic field at that point. The lines connecting the direction of the field are called *lines* of *force*. The lines of force existing around a bar magnet are shown in Fig. 9–1.

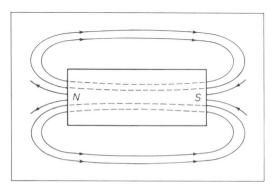

figure 9–1
magnetic field around a bar magnet

magnetic fields

A Danish physicist, Oersted, discovered in the early nineteenth century that a current-carrying wire is surrounded by a magnetic field. The lines of force about a current-carrying conductor can be clockwise or counter-clockwise in direction, depending on the direction of the electric current flow. It should be noted that magnetic lines of force always form complete loops.

Although the magnetic field around a straight current-carrying wire has no north or south poles, the lines of force have direction. This direction may be determined by the *right-hand rule*, as shown in Fig. 9–2. For the application of the right-hand rule, assume that you are holding the conductor with the right hand in such a manner that the thumb points in the direction of conventional current flow; then the fingers will indicate the direction of the line of force.

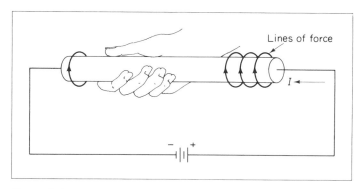

figure 9–2
right-hand rule for determination of direction

If several turns of wire are wound close together to form a coil, the magnetic fields about each turn will all have the same direction. A coil of this type is called a solenoid and is shown in Fig. 9–3.

The right-hand rule can be applied to the solenoid in the following manner. Assume the thumb of the right hand to be pointing in the conventional current flow direction; the fingers will indicate the direction of the magnetic lines of force.

The density of the magnetic lines of force can be increased by inserting a piece of iron, steel, or cobalt within the solenoid. By this method, we have created an electromagnet and effectively increased both the flux density and the applied field strength.

The unit for magnetic lines of force or flux in the MKS system is called the *weber*. The letter symbol for magnetic flux is the Greek letter φ (phi).

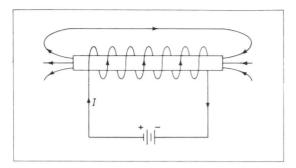

figure 9–3
right-hand rule. applied to solenoid

magnetomotive force

Analogous to the flow of electric current when an emf source is applied, magnetic flux does not exist until a magnetizing force is established. It should be noted that magnetic flux does not appear until electric current starts to flow in the solenoid. Consequently, the magnetomotive force must be proportional in some manner to the electric current. In general, the magnetomotive force established is directly proportional to the electric current and also to the number of turns of wire in the coil. Thus,

$$\mathscr{F} = NI \text{ (ampere turns)} \tag{9–1}$$

where

I = current through the wire in amperes.
N = number of turns in the coil.
$\mathscr{F}$ = magnetomotive force in ampere turns.

Note that the number of ampere turns is the MKS unit for magneto-motive force. The letter symbol is the script letter $\mathscr{F}$ (eff).

reluctance

The reluctance of a magnetic circuit is analogous to the resistance of an electric circuit. The Ohm's law analogy can be applied directly to the magnetic circuit as

$$\mathscr{R} = \frac{\mathscr{F}}{\varphi} \text{ rels or } \frac{\text{ampere turns}}{\text{weber}} \tag{9–2}$$

The script letter $\mathscr{R}$ (ar) represents the symbol for reluctance in a magnetic circuit.

permeability

In the analysis of parallel circuits, it was convenient sometimes to consider the conductance of the circuit elements rather than their resistance. In dealing with magnetic circuits it might be simpler to measure the ability of a magnetic circuit to permit the establishment of magnetic flux. This ability is called the *permeance* of a magnetic circuit.

A comparison of the magnetic properties of various materials indicates a direct proportion to the unit length and an inverse proportion to the cross-sectional area of the element. This, of course, is analogous to the permittivity of an electric circuit element.

The permeance, as a function of a unit length and a cross-sectional area, is called the permeability. The letter symbol for permeability is the Greek letter μ (mu). The mathematical definition for permeability is

$$\mu = \frac{l}{\mathscr{R}A} \qquad (9\text{--}3)$$

where

μ = permeability in webers/ampere turn/meter.
l = length of magnetic circuit in meters.
A = cross-sectional area in square meters.
$\mathscr{R}$ = reluctance in ampere turns/weber.

flux density and magnetizing force

The magnetic flux per unit area is defined as the flux density and is symbolized by the capital letter B. The mathematical relationship for flux density is

$$B = \frac{\varphi}{A} \frac{\text{webers}}{\text{square meter}} \qquad (9\text{--}4)$$

It is evident that the unit for flux density is webers per square meter. Note also that the permeability deals with the volume of a magnetic material. Consequently, the magnetic unit of interest is the magnetomotive force per unit length, called the *magnetizing force*. The letter symbol for the magnetizing force is the capital letter H. The mathematical relationship for magnetizing force is

$$H = \frac{\mathscr{F}}{l} \quad \frac{\text{ampere turn}}{\text{meter}}$$

(9–5)

The interrelationship between B and H can be determined by

$$\mu = \frac{l}{\mathscr{R}A} = \frac{l}{\left(\dfrac{\mathscr{F}}{\varphi}\right)A} = \frac{\dfrac{\varphi}{A}}{\dfrac{\mathscr{F}}{l}}$$

$$\mu = \frac{B}{H}$$

(9–6)

magnetism

Magnetic field lines are unaffected by nonmagnetic materials such as paper, wood, air, etc. The permeability of free space is equal to

$$\mu_0 = 4\pi \times 10^{-7}$$

(9–7)

Some elements have a permeability slightly less than that of free space. Such elements are said to be diamagnetic materials. A few materials have a permeability slightly greater than that of free space and are said to be paramagnetic materials. Some examples of paramagnetic materials are platinum, aluminum, and oxygen. Elements that have extremely high permeabilities compared to the permeability of free space are said to be ferromagnetic materials.

In an unmagnetized state, the ferromagnetic material has a net mmf equal to zero. When an external magnetizing force is applied, the electron orbits of the atoms align themselves to form a net magnetic field in one direction, as shown in Fig. 9–4.

This alignment process is called *magnetization* of the material. The grouping of a specified number (between 10^{12} and 10^{15}) of atoms within the material forms a *domain* region. Each domain is separate and independent of the surrounding domains. Each domain is considered a small bar magnet generating lines of flux that become additive and increase the magnetic flux generated by the electromagnet. When all the domains have

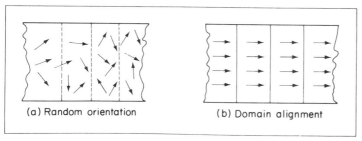

figure 9–4
magnetism

been aligned, the ferromagnetic material is considered to be in saturation. A further increase in magnetizing force will have no effect on the domains within the material.

hysteresis curves

A graph indicating the relationship between B (flux density) and H (magnetizing force) is usually presented by the manufacturer of magnetic materials to the practicing engineer. It should be noted that the magnetizing force can be controlled by varying the electric current through a solenoid.

Consider the magnetic circuit shown in Fig. 9–5. Assume that initially the iron is unmagnetized with current $I = 0$. The application of voltage E causes current I to flow. The magnetization force H is equal to ampere turns per unit length. The flux density B is defined as the magnetic flux per unit area and will also increase with an increase in the current I. Consequently, the B–H curve produced will vary according to the graph shown in Fig.

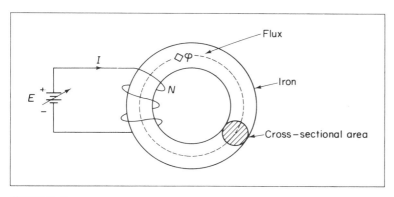

figure 9–5
magnetic circuit used for B–H curve

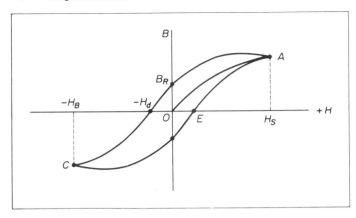

figure 9–6
typical hysteresis loop

9–6. Figure 9–6 shows the complete cycle of magnetization of a piece of iron or any ferromagnetic material. The procedure is presented in a step by step manner.

step 1

Starting from zero, an increase in current causes the flux density and the magnetizing force to reach point A along the path shown in Fig. 9–6.

step 2

The values of B and H at point A are considered the saturation values. A further increase in I will not produce a larger value of B. As current I is reduced toward zero, the flux density decreases from saturation to a point designated as B_R. B_R is called the residual flux density.

step 3

At this point, the current in the circuit is zero. In order to bring the flux density within the iron to zero, the source emf must reverse to cause current I to reverse; current then increases in the negative direction to $-H_d$.

step 4

H_d is called the demagnetizing force required to bring the residual flux density to zero. A further increase in current in the negative direction will result in saturation at point C.

step 5

Again, reducing the current I towards zero causes the curve to reach point D.

step 6

The source voltage E must be reversed again to increase the current I in the positive direction. This causes point D to move to point E. A further increase in current I to maximum completes the cycle and moves point E to point A.

It should be noted that the flux density tends to lag behind the magnetizing force that created it. The cyclical curve is called the hysteresis loop for a specified magnetic material.

The area within the hysteresis loop is equal to the product of B and H and is measured in joules per square meter. It should be noted that an increase in the residual flux density B_R of a magnetic material results in an increased area within the hysteresis loop. Consequently, the area within the hysteresis loop is a means of comparing the permeabilities of various magnetic materials.

Typical $B-H$ curves for various materials are shown in Fig. 9–7.

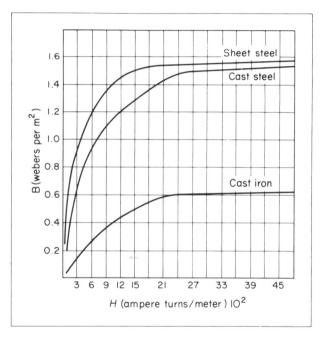

figure 9–7
normal magnetization curves

magnetic circuits (series)

The Ohm's law relationship for electric circuits is that the resistance of the circuit is equal to the voltage-to-current ratio. The analogous magnetic circuit relationship is

$$\mathscr{R} = \frac{\mathscr{F}}{\varphi}$$

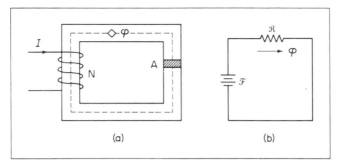

figure 9–8
(a) simple magnetic circuit; (b) equivalent circuit

The Kirchhoff's voltage law relationship for an electric circuit has a magnetic circuit analogy, which is

$$\Sigma_\emptyset \mathscr{F} = 0$$

Consider the magnetic circuit shown in Fig. 9–8. A coil of N turns is wound on a ferromagnetic core. The length of the flux path is defined as l. The cross-sectional area of the core is constant throughout and equal to A. Note that the mmf is equal to

$$\mathscr{F} = \mathscr{R}\varphi$$

Attention must be given to the magnetic circuit to determine the correct value of the length and cross-sectional areas of magnetic circuits. Assume a magnetic circuit with two different cross-sectional areas as shown in Fig. 9–9.

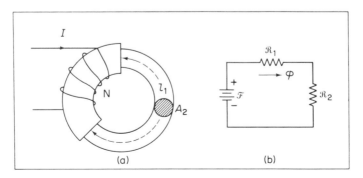

figure 9–9
(a) actual magnetic circuit; (b) equivalent circuit

Given the circuit shown in Fig. 9–9 with the following given values,

$$A_1 = 9 \text{ m}^2 \qquad l_1 = 5 \text{ m} \qquad N = 50$$
$$A_2 = 6 \text{ m}^2 \qquad l_2 = 50 \text{ m} \qquad I = 1 \text{ A}$$
$$\mu = 8.5 \times 10^{-3}$$

find the total flux in the coil.

solution

step 1: Solve for $\mathscr{R}_1$.

$$\mathscr{R}_1 = \frac{l_1}{\mu A_1} = \frac{5}{8.5 \times 10^{-3} \times 9}$$

$$\mathscr{R}_1 = 65.5 \text{ rels}$$

step 2: Solve for reluctance $\mathscr{R}_2$.

$$\mathscr{R}_2 = \frac{l_2}{\mu A_2} = \frac{50}{8.5 \times 10^{-3} \times 6}$$

$$\mathscr{R}_2 = 980.5$$

step 3: Calculate $\mathscr{F}$.

$$\mathscr{F} = NI = 50 \times 1 = 50 \text{ AT (ampere turns)}$$

step 4: Calculate φ.

$$\varphi = \frac{\mathscr{F}}{\mathscr{R}} = \frac{50}{980.5 + 65.5}$$

$$\varphi = 47.8 \times 10^{-3} \text{ webers}$$

For the following series magnetic circuit, find the value of current required to produce a magnetic flux of 3×10^{-3} weber.

Section	Material	Length	Area
a–b	cast steel	12 cm	6×10^{-3} m²
b–c	cast iron	12 cm	6×10^{-3} m²
a–c	sheet steel	12 cm	6×10^{-3} m²

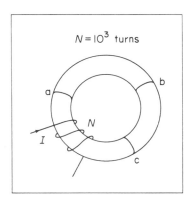

$N = 10^3$ turns

solution

step 1: Construct the equivalent circuit. There are essentially three reluctances and consequently three mmf values to be calculated.

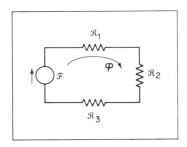

step 2: Calculate B_1 for section a–b.

$$B_1 = \frac{\varphi}{A} = \frac{3 \times 10^{-3}}{6 \times 10^{-3}} = 0.5 \text{ weber/m}^2$$

From curves,

$H_1 = 300$ ampere turns/meter

step 3: Calculate B_2 for section b–c. Since φ and A are constant, B must
also be constant. Thus,

$$B_2 = 0.5 \text{ w/m}^2$$

From curves,

$H_2 = 1500$ AT/m

step 4 :   Calculate H_3 for section a–c. From curves,

$$H_3 = 50 \text{ AT/m}$$

step 5 :   Sum up the total ampere turns.

$$H_1 l = 300 \times 0.12 = 36$$
$$H_2 l = 1500 \times 0.12 = 180$$
$$H_3 l = 50 \times 0.12 = 6$$

Total = 222 AT

step 6 :   Calculate I.

$$= NI = 222 \text{ AT}$$

$$I = \frac{222}{1000} = 222 \text{ ma}$$

---

sample problem

The following magnetic circuit with two input sources has the following given data.

$A = 10 \text{ cm}^2$ $N_1 = 1000$ turns
$l_1 = 12 \text{ cm}$ $N_2 = 3000$ turns
$l_2 = 6 \text{ cm}$ $I_2 = 200 \text{ ma}$

Find the current I_1 to produce a flux $\varphi = 1.2 \times 10^{-4}$ weber.

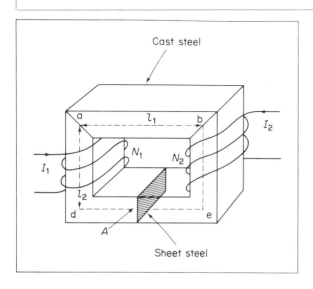

Cast steel

Sheet steel

solution

step 1: The equivalent circuit is

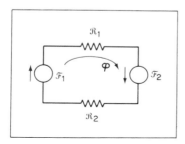

The two reluctances are unknown.
It usually is desirable to set up a table as follows:

Section	Material	l	A (m²)	B	H
a–b	cast steel	0.12 m	10^{-3}		
$bcda$	sheet steel	0.24 m	10^{-3}		

step 2: Calculate B_1.

$$B_1 = \frac{\varphi}{A} = B_2 = 1.2$$

From curves for cast steel,

$$H = 1150 \text{ AT/m}$$

From curves for sheet steel,

$$H = 375 \text{ AT/m}$$

step 3: Calculate the total ampere turns.

$$Hl_1 = 1150 \times 0.12 = 1380 \text{ AT} \quad \text{(cast steel)}$$
$$(l_1 + 2l_2)H = 375 \times 0.24 = 90 \text{ AT} \quad \text{(sheet steel)}$$

step 4: Calculate I_1.

$$\mathscr{F}_1 + \mathscr{F}_2 = 1470 \text{ AT}$$
$$1000 \, I_1 = \mathscr{F}_1$$
$$600 = \mathscr{F}_2$$

Then

$$I_1 = \frac{870}{1000} = 870 \text{ ma}$$

air gaps

In the magnetic circuits under consideration, a complete closed path was presented to the flux. Consider the effects of an air gap within a magnetic circuit as shown in Fig. 9–10. Every line of magnetic flux must cross the air gap. The spreading of the lines of flux outside the common area of the core for the air gap is known as *fringing*. The air gap is necessary in electrical systems to provide room for electrical conductors to rotate or move.

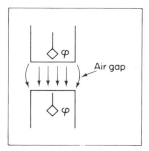

figure 9–10
fringing effects in an air gap

An air gap is in series with the remainder of the circuit. Consequently, all magnetic circuits with gaps are solved in a similar manner as if all the different elements were in series with respect to the input magnetomotive force.

The flux density of the air gap is determined by

$$B_a = \frac{\varphi}{A_a}$$

In a series magnetic circuit, the lines of flux are constant to all the elements.

The permeability of air is equal to that of free space. The magnetizing force is then given by

$$H_a = \frac{B_a}{\mu_0}$$

where

$$\mu_0 = 4\pi \times 10^{-7}$$

An illustrative problem will demonstrate the theory.

sample problem

Determine the value of I required to establish a flux of 4×10^{-4} weber. The following data are specified.

Section	Material	l	A	B	H
a–b	sheet steel	6 cm	4.5 cm²		
b–c	air gap	0.1 mm	4.5 cm²		
c–d	sheet steel	6 cm	4.5 cm²		
d–a	cast steel	12 cm	4.5 cm²		

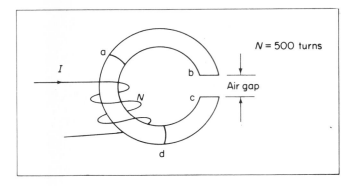

solution

step 1: The resultant equivalent circuit is

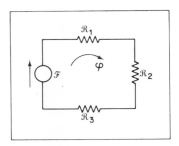

The flux density is constant throughout. Thus,

$$B = \frac{\varphi}{A} = \frac{4 \times 10^{-4}}{4.5 \times 10^{-4}} = 0.89 \text{ weber/m}^2$$

step 2: Determine the values of H. From the curves,

Sheet steel:

$H = 250 \text{ AT/m}$

Cast steel:

$H = 590$ AT/m

Air gap:

$H = \dfrac{B}{\mu_0} = 70.7 \times 10^4$ AT/m

step 3: Evaluate the total ampere turns.

Sheet steel:

$Hl_1 = 250 \times 0.12 = 30$ AT

Cast steel:

$Hl_2 = 590 \times 0.12 = 70.8$ AT

Air gap:

$Hl_3 = 70.7 \times 10^4 \times 10^{-4} = 70.7$ AT

Total AT $= 171.5$ AT

step 4: Calculate I.

$\mathscr{F} = NI = 171.5$ AT

$I = \dfrac{171.5}{500} = 343$ ma

parallel magnetic circuits

A parallel electric circuit has a number of resistors connected across the source. An analogous magnetic circuit would have a number of reluctances connected across the source or magnetomotive force.

Consider the magnetic circuit shown in Fig. 9–11. If the two parallel

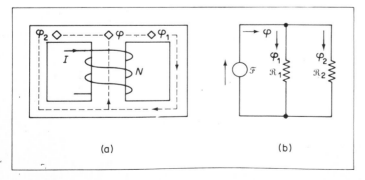

(a) (b)

figure 9–11
(a) parallel magnetic circuit; (b) equivalent circuit

paths are identical, then $\varphi_1 = \varphi_2 = \varphi/2$. Two equal parallel paths cause the total flux to split into equal halves.

An illustrative problem will demonstrate the theory.

---

sample problem

The following magnetic circuit and data are given. The material of the core is sheet steel. The number of turns total 500. The length of the mean path is 25 cm. The thickness is 5 cm throughout. The width of the outer core is 3 cm, whereas the width of the inner core is 6 cm. Find the current required to produce a total flux of 20×10^{-4} weber.

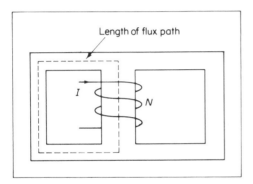

Length of flux path

solution

step 1: Since the two halves are equal, we can consider one side of the magnetic circuit and assume that it conducts half of the total flux. The cross-sectional area is

$$A = 3 \times 5 = 15 \text{ cm}^2$$

step 2: Calculate B and H.

$$B = \frac{\varphi}{A} = \frac{10 \times 10^{-4}}{15 \times 10^{-4}} = 0.667$$

From curves for sheet steel,
$$H = 300 \text{ AT/m}$$

step 3: Calculate the magnetomotive force.

$$\mathscr{F} = Hl = 300 \, (0.25) = 75 \text{ AT}$$

step 4: Calculate I.

$$I = \frac{\mathscr{F}}{N} = \frac{75}{500} = 150 \text{ ma}$$

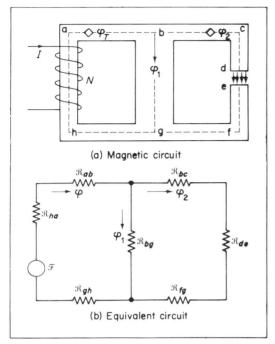

(a) Magnetic circuit

(b) Equivalent circuit

figure 9–12
series-parallel magnetic circuit

It is evident that combinations of series and parallel magnetic circuits exist. For example, note the series-parallel magnetic circuit shown in Fig. 9–12.

problems

1. Given the electromagnet shown, determine the flux density of the core.

$\varphi = 8 \times 10^{-4}$ weber
$A = 0.05$ m^2

2. Find the reluctance of a magnetic circuit if the magnetic flux φ is equal to 5×10^{-4} weber and is created by an mmf of 600 AT.

3. Find the reluctance of a magnetic circuit if the mmf applied is 300 AT and establishes a magnetic flux of 7.5×10^{-4} weber.

4. If a magnetizing force of 600 AT/m is applied and the flux density is 0.12 weber per m², determine (a) the permeability of the material, (b) the permeability of the material if the flux density is doubled and the magnetizing force remains constant.

5. The cast steel core shown has the following dimensions.
$d_i = 8$ m, $A = 12$ cm², $d_o = 12$ m, $N = 2000$
Determine the current I required to produce a magnetic flux of 10^{-3} weber.

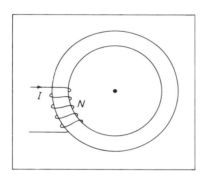

width = 4 cm   thickness = 1.5 cm $N = 250$ turns

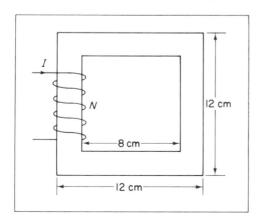

diagram for problems 6, 7, 8, and 9

6. Determine the current required to establish a magnetic flux of 0.9×10^{-3} weber through the sheet steel core.

7. Determine the current required to establish a magnetic flux of 0.6×10^{-3} weber through the cast steel core.

8. Consider the circuit of Problem 6 having an unknown number of turns after the 250 turns have been removed. Determine the number of turns required if the input current of 200 ma establishes a flux of 0.75×10^{-3} weber through the sheet steel core.

9. Remove the 250 turns from the cast steel core of Problem 7. Determine the number of turns required if an input current of 500 ma establishes a flux of 0.75×10^{-3} weber.

10. What current must flow in the given circuit to produce a flux of 10×10^{-4} weber?

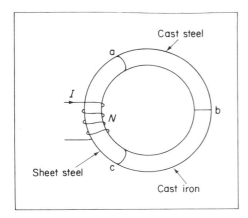

$$N = 100 \qquad m^2$$
$$l_{a-b} = 5 \text{ cm} \qquad A_{a-b} = 2.5 \times 10^{-3}$$
$$l_{b-c} = 5 \text{ cm} \qquad A_{b-c} = 2.5 \times 10^{-3}$$
$$l_{a-c} = 5 \text{ cm} \qquad A_{a-c} = 2.5 \times 10^{-3}$$

11. Find the number of turns required to establish a flux of 10^{-3} weber in a cast steel core. (See figure on next page.)
$$I_1 = 1 \text{ A,} \qquad I_2 = 1.5 \text{ A,} \qquad N_1 = 100 \text{ turns}$$

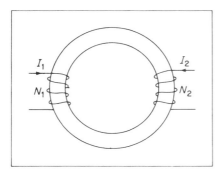

$A = 12$ cm²
$l = 20$ cm

diagram for problems 11, 12, 13, and 14

12. Find the number of turns required to establish a flux of 10^{-3} weber in a sheet steel core.
$I_1 = 1$ A, $I_2 = 2$ A, $N_2 = 100$ turns

13. Find the current required to establish a flux of 6×10^{-4} weber in a cast steel core.
$I_1 = 2$ A, $N_1 = 50$ turns, $N_2 = 100$ turns

14. Find the current required to establish a flux of 6×10^{-4} weber in a sheet steel core.
$I_2 = 0.12$ A, $N_1 = 100$ turns, $N_2 = 50$ turns

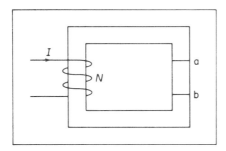

diagram for problems 15, 16, 17, and 18

15. Determine the current required to establish a flux of 8×10^{-4} weber. Material from a–b is cast steel. Uniform area $= 16$ cm². All other material is sheet steel. Mean length of path a–$b = 2$ cm; other $= 18$ cm; $N = 200$ turns.

16. Determine the current required to establish a flux of 8×10^{-4} weber. Material from a–b is sheet steel. Uniform area $= 16$ cm². All other material is cast steel. Mean length of path a–$b = 2$ cm; other $= 18$ cm; $N = 200$ turns.

17. Determine the current required to establish a flux of 12×10^{-4} weber. Material from a–b is cast steel. Uniform area = 20 cm². All other core material is sheet steel. Mean length of path a–b = 4 cm; other = 16 cm; N = 200 turns.

18. Determine the current required to establish a flux of 12×10^{-4} weber. Material from a–b is sheet steel. There is a uniform area throughout of 20 cm². All other core material is cast steel. Mean length of path a–b = 4 cm; other = 16 cm; N = 200 turns.

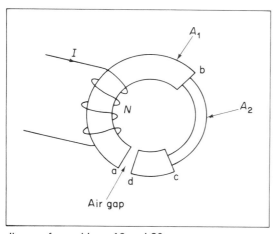

diagram for problems 19 and 20

19. Determine the current required to establish a flux of 8×10^{-4} weber. Assume all flux lines through the core.

$A_1(a$–$b) = 24$ cm², $A_2 = 10$ cm²
$A_1(a$–$d) = 24$ cm², $N = 500$ turns

Material	Section	Length
Sheet steel	a–b	15 cm
Cast steel	b–c	10 cm
Sheet steel	c–d	5 cm
Air	d–a	0.1 mm

20. Determine the current required to establish a flux of 8×10^{-4} weber. Assume that all flux passes through the core. N = 500 turns.

Section	Material	Length	Area
a–b	cast steel	15 cm	24 cm²
b–c	sheet steel	10 cm	10 cm²
c–d	cast steel	5 cm	24 cm²
d–a	air	0.1 mm	24 cm²

10

inductors

introduction

Basically, there are three properties that circuit elements can have. A circuit can have *resistance*, *capacitance*, and *inductance*.

Heretofore, the magnitude of current that flows when circuit elements were connected to a source depended on the resistance of the devices. Consequently, a *resistor* was defined as that property of a circuit element that opposes the flow of current. A second property of a circuit element, *capacitance*, opposes any change in the voltage across it. A third property of a circuit element, *inductance*, opposes any change in the current that flows through it. Just as the *capacitance* of a circuit is associated with the *electric field* accompanying a difference of potential, the *inductance* of an electric circuit is associated with the *magnetic field* around a current-carrying conductor.

Faraday's law

Consider the situation of a moving conductor cutting magnetic lines of flux as shown in Fig. 10–1. As the conductor is moved through the magnetic field so that it cuts magnetic lines of flux, the needle pointer of the galvonometer registers a finite value of current. When there is no motion of the conductor, the needle registers or remains at zero. Consequently, for current to flow through the galvanometer a source emf in the closed loop must exist. Further experiment indicates that moving the conductor in a motion parallel to the magnetic lines of force has no effect on the galvonometer.

The source emf appears only when the conductor is cutting across the magnetic lines of force. The same effect can also be accomplished by maintaining the conductor fixed in position and moving the magnetic lines of force in a vertical manner. The generation of an emf by the cutting of magnetic lines of force by a conductor is called *electromagnetic induction*, and the emf produced or induced is called an *induced emf*.

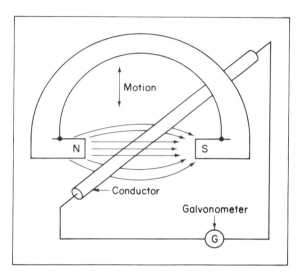

figure 10–1
electromagnetic induction

If a coil of N turns is placed across a varying magnetic field, a voltage will be induced across the coil as determined by Faraday's law.

$$e = N \frac{d\varphi}{dt} \text{ volts} \tag{10-1}$$

where

N = number of turns on the coil.

$\dfrac{d\varphi}{dt}$ = time rate change in flux (webers/sec).

Since a conductor cannot store electric charge, the emf is existent only while the motion of the conductor or magnetic field is producing the induced emf.

Lenz's law

The concept has been developed that the motion of magnetic flux cutting perpendicularly across a conductor generates an induced current in the conductor. The electromagnetic induction is increased if a coil is used as the conductor. If the current input to a coil is increased, the flux linkages increase with a consequent increase in induced voltage. For the coil, therefore, an induced voltage is developed across the coil due to the time rate change of current. A statement of Lenz's law is: The polarity of this induced emf must be such as to oppose the cause that produces it.

Consider the circuit shown in Fig. 10–2. The motion of the conductor across a magnetic flux induces an emf in the conductor with the polarity indicated. This emf produces a current in the loop flowing in the conventional current direction.

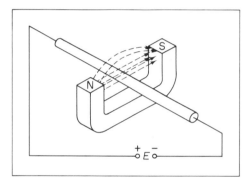

figure 10–2
Lenz's law application

self-induction

According to Lezn's law, the voltage induced in a conductor or coil opposes any change in current flowing through the coil. This induced voltage is called a counter emf. The effectiveness with which a coil opposes any time rate change of current is a property called the *inductance* of a coil. The induction of an emf by a time rate change of current in the same circuit is called *self-induction*.

The letter symbol for inductance is L. It should be noted that the induced voltage of a coil is directly proportional to both the inductance and the time rate change of current. Consequently, the mathematical relationship is

$$e = L \frac{di}{dt} \text{ volts} \tag{10–2}$$

The basic unit of inductance is the henry. A coil has an inductance of 1 henry when the time rate change of current of 1 ampere per second produces a counter emf of 1 volt in that coil. The symbol for inductance is shown in Fig. 10–3.

The evaluation of the inductance of a coil as a function of the number of turns and coil geometry can be derived in the following manner. Both Faraday's law and Lenz's law deal with the equation for the self-induced voltage within a coil. Setting these two equations equal yields

$$L\frac{di}{dt} = N\frac{d\varphi}{dt}$$

and

$$L = N\frac{d\varphi}{di}$$

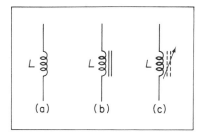

figure 10–3
(a) air core; (b) iron core; (c) variable tuned

It can be shown that

$$\frac{d\varphi}{di} = N\frac{\mu A}{l}$$

Substituting results in

$$L = N^2\frac{\mu A}{l}\ H \qquad (10\text{–}3)$$

The inductance of a wire conductor or coil is directly proportional to the square of the number of turns, the relative permeability of the core material, and the cross-sectional area of the material in square meters and inversely proportional to the length of the magnetic path in meters. An illustrative problem will demonstrate the theory.

---

sample problem

Given the circuit shown, determine the coil inductance.

$d = 4 \times 10^{-3} \text{m}$ $l = 20 \text{ cm}$ $N = 200 \text{ turns}$

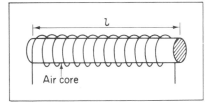

Air core

solution

step 1:   Calculate the area A.

$$A = \frac{\pi \, d^2}{4} = \frac{\pi(4 \times 10^{-3})^2}{4}$$

$$A = 12.56 \times 10^{-6} \text{ m}^2$$

step 2:   Calculate L.

$$L = N^2 \frac{\mu A}{l} = (200)^2 \frac{4\pi \times 10^{-7} \times 12.56 \times 10^{-6}}{0.2}$$

$$L = 31.4 \ \mu H$$

In applying the equation for the inductance of a coil, it should be noted that an assumption is made that all of the flux generated links all of the turns. This means that fringing effects and other leakage sources are ignored. This equation is fairly accurate for iron coils, but for air core coils, experimental procedures and formulas derived in field practice are used. These formulas can be found in various electrical and electronic handbooks.

inductances in series and parallel

When two inductors are connected in series as shown in Fig. 10–4, the same current flows in both. Assume that the two inductors are connected physically in such a manner that the magnetic field of one cannot induce an

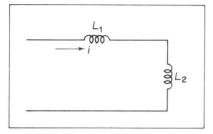

figure 10–4
two inductors in series

emf in the other coil. When inductors are in series, the total inductance of the circuit can be derived with a general approach by

$$E = e_1 + e_2 + e_3 + \cdots + e_n$$

Since

$$L_T \frac{di}{dt} = L_1 \frac{di}{dt} + L_2 \frac{di}{dt} + L_3 \frac{di}{dt} + \cdots + L_n \frac{di}{dt}$$

then

$$L_T = L_1 + L_2 + L_3 + \cdots + L_n \qquad (10\text{–}4)$$

Therefore, when inductors are connected in series, the total inductance is equal to the sum of the individual inductances, just like resistors in series.

When two inductors are connected in parallel as shown in Fig. 10–5, the total inductance can be derived as follows.

$$E = L_1 \frac{di_1}{dt} + L_2 \frac{di_2}{dt} + L_3 \frac{di_3}{dt} + \cdots + L_n \frac{di_n}{dt}$$

and

$$\frac{E}{L_T} = \frac{E}{L_1} + \frac{E}{L_2} + \frac{E}{L_3} + \cdots + \frac{E}{L_n}$$

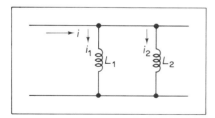

figure 10–5
two inductors connected in parallel

and

$$\frac{1}{L_T} = \frac{1}{L_1} + \frac{1}{L_2} + \frac{1}{L_3} + \cdots + \frac{1}{L_n}$$

(10–5)

It is evident that inductors connected in parallel are treated just like resistors in parallel.

RL circuit

Consider the series RL circuit shown in Fig. 10–6. When the switch is open or in position 1, the current in the circuit is zero. At the instant

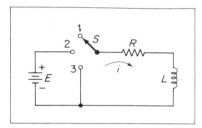

figure 10–6
series RL circuit

the switch is closed to position 2, the inductor acts to prevent the instantaneous flow of current. Since the current in the circuit is zero, the potential drop across the coil is equal to the applied voltage. The current i will start to increase from zero with a consequent drop in voltage across R and a decrease in voltage across the inductor L. Eventually, the voltage across the inductor will be zero, while that across the resistor will be equal to the applied voltage.

analysis

The nomenclature that will be used in this section is

(a) All lowercase letters will represent instantaneous values that are functions of time.

(b) Uppercase letters will represent steady-state conditions.

The application of Kirchhoff's law equation yields the following resultant equation:

$$E = iR + L\frac{di}{dt}$$

Note that at $t = 0$, $i = 0$; then

$$\frac{di}{dt} = \frac{E}{L} \quad (\text{at } t = 0)$$

The solution for current in the circuit if calculus is used is

$$i = \frac{E}{R}(1 - \epsilon^{-Rt/L}) \text{ amperes} \qquad (10\text{--}6)$$

This equation is analogous to the equation for the charging voltage across a capacitor and has a graph like that shown in Fig. 10–7.

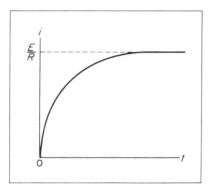

figure 10–7
variation of current with respect to time

Note that for capacitive circuits the product of RC defined the time constant, whereas, for RL circuits, the time constant is equal to the ratio of L to R. Thus,

$$\tau = \frac{L}{R} \text{ seconds} \qquad (10\text{--}7)$$

where

$L =$ inductance in henrys.
$R =$ resistance in ohms.

The time constant τ determines the rapidity with which the curves rise from zero to the final steady-state value of current of E/R. The voltage across the resistor is

$$e_R = iR = E(1 - \epsilon^{-t/\tau}) \qquad (10\text{–}8)$$

The voltage across the inductor L is

$$e_L = E - e_R = E\epsilon^{-t/\tau} \text{ volts} \qquad (10\text{–}9)$$

An illustrative problem will demonstrate the theory.

sample problem

Given the following circuit and data, determine, at $t = 6$ ms, (a) the instantaneous current; (b) the voltage across R; (c) the voltage across L.

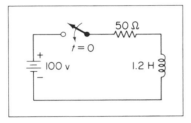

solution

step 1: Calculate τ.

$$\tau = \frac{L}{R} = \frac{1.2}{50}$$

$$\tau = 24 \text{ ms}$$

step 2: Calculate i.

$$i = \frac{E}{R}(1 - \epsilon^{-t/\tau})$$

$$i = \frac{100}{50}(1 - \epsilon^{-0.25})$$

$$i = 2(1 - 0.78) = 0.44 \text{ ampere}$$

step 3: Calculate e_R.

$$e_R = iR = 0.44 \times 50 = 22 \text{ volts}$$

step 4:   Calculate e_L.

$$e_L = E - e_R = 78 \text{ volts}$$

sample problem

Given the following circuit and data, determine, after 0.6 μsec, (a) the instantaneous current; (b) the voltage across R; (c) the voltage across L.

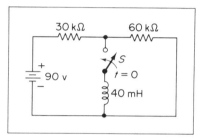

solution

step 1:   Thevenize the circuit.

$$R_{th} = \frac{30 \times 60}{90} 10^3 = 20 \text{ k}\Omega$$

$$E_{oc} = \frac{90}{(30 + 60)} 60 = 60 \text{ v}$$

The new equivalent circuit is

step 2:   Calculate the time constant.

$$\tau = \frac{L}{R} = \frac{40 \times 10^{-3}}{20 \times 10^3} = 2 \text{ } \mu\text{sec}$$

step 3:   Calculate i.

$$i = \frac{E}{R}(1 - \epsilon^{-t/\tau})$$

$$i = \tfrac{60}{20}(1 - \epsilon^{-0.3})\ 10^{-3}$$

$$i = 3(1 - 0.74)\ \text{ma}$$

$$i = 780\ \mu\text{a}$$

step 4:   Calculate e_R.

$$e_R = iR = 0.78 \times 10^{-3} \times 20 \times 10^{3}$$

$$e_R = 15.6\ \text{v}$$

step 5:   Calculate e_L.

$$e_L = E - e_R = 44.4\ \text{v}$$

energy stored by an inductance

The ideal inductor, like the capacitor, does not dissipate any electrical energy, but stores it in the magnetic field. The source emf is supplying energy to the inductor at the rate of $p_L = e_L i$.

The energy stored by an inductor can be determined by choosing some final value of current and evaluating the area under the power curve. Figure 10–8 shows the graphical representation of instantaneous power curve. Evaluation of the area under the power curve yields the energy stored in the magnetic field of the inductor.

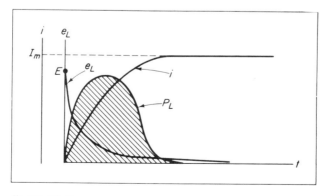

figure 10–8
graphical representation of the power curve in an inductor

The mathematical analysis for the energy stored is

$$w = \int_0^t p_L dt = \int_0^t e_L i \, dt$$

Substituting for e_L yields

$$w = \int_0^t Li \frac{di}{dt} dt = \int_0^t Li \, di$$

$$\boxed{w = \tfrac{1}{2} Li^2 \text{ joules}} \tag{10-10}$$

It is evident that the energy stored in an inductor at any instant of time is directly proportional to both the inductance and the square of the instantaneous current flowing through it. When current reaches steady-state value, the energy stored in the inductor is

$$\boxed{W = \tfrac{1}{2} LI^2 \text{ joules}} \tag{10-11}$$

where $I = E/R$.

RL discharging circuits

The diagram shown in Fig. 10–6 can also illustrate the decaying action of an RL series circuit. When the switch is in position 2, a steady-state current equal to E/R eventually will flow. When the switch is moved to position 3, the energy stored in the magnetic field of the inductor tends to maintain the current flow in the same direction through the resistor. After some time has elapsed, the dissipated energy across the resistor will cause the stored energy to decrease to zero, and the current in the circuit falls to zero.

The mathematical analysis utilizes the Kirchhoff's law equation. Thus,

$$Ri + L \frac{di}{dt} = 0$$

Solving this equation by the use of calculus yields

$$i = I \, \epsilon^{-t/\tau} \tag{10-12}$$

where

$$I = \frac{E}{R}.$$

$$\tau = \frac{L}{R}.$$

The voltage across the resistor is

$$e_R = iR = E \, \epsilon^{-t/\tau} \qquad\qquad (10\text{–}13)$$

The voltage across the inductor is

$$e_L = -e_R = -E \, \epsilon^{-t/\tau} \qquad\qquad (10\text{–}14)$$

A graph of current and voltages in an *RL* discharge circuit is shown in Fig. 10–9.

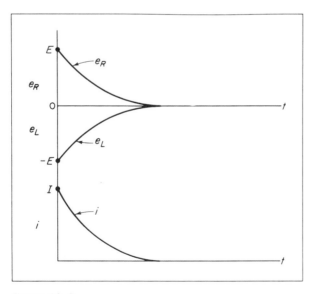

figure 10–9
discharge curves for an *RL* circuit

Note that in the instantaneous discharge graph shown in Fig. 10–9 the curves all follow the basic exponential characteristic. A universal time constant graph is shown in Fig. 10–10.

An illustrative problem will demonstrate the theory.

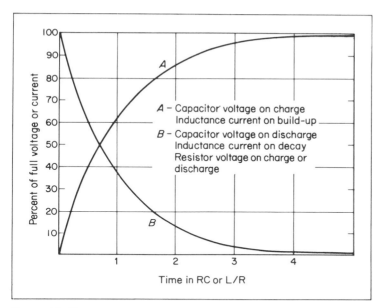

figure 10–10
universal time constant chart

---

sample problem

The switch in the circuit shown is closed for 9 μsec and then opened.
Determine, at $t = 4$ μsec, (a) the instantaneous current; (b) the voltage
across R; (c) the voltage across L.
At 3.6 μsec after the switch is opened, determine (d) the instantaneous
current; (e) the voltage across R; (f) the voltage across L.

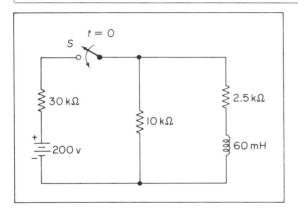

solution

step 1: Thevenize the circuit.

$$R_{th} = \frac{10 \times 30}{10 + 30} \, 10^3 = 7.5 \text{ k}\Omega$$

$$E_{oc} = 200 \, \frac{10}{10 + 30} = 50 \text{ v}$$

The new equivalent circuit is

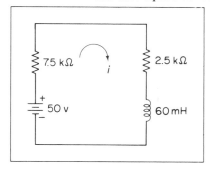

step 2: Calculate the time constant τ.

$$\tau = \frac{L}{R} = \frac{60 \times 10^{-3}}{10 \times 10^3} = 6 \text{ } \mu\text{sec}$$

step 3: Calculate i.

$$i = \frac{E}{R} (1 - \epsilon^{-t/\tau}) \text{ A}$$

$$i = \tfrac{50}{10} (1 - \epsilon^{-4/6}) \text{ ma}$$

$$i = 5(1 - 0.51) \text{ ma}$$

$$i = 2.45 \text{ ma}$$

step 4: Calculate e_R.

$$e_R = iR = 2.45 \times 10 = 24.5 \text{ v}$$

step 5: Calculate e_L.

$$e_L = E - e_R = 50 - 24.5$$

$$e_L = 25.5 \text{ v}$$

step 6: Calculate the current at 9 μsec.

$$i = \frac{E}{R} (1 - \epsilon^{-9/6}) = 5(1 - 0.22)$$

$$i = 3.9 \text{ ma}$$

step 7: The new equivalent circuit after the switch is open is

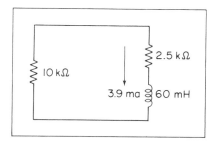

Calculate the new time constant.

$$\tau = \frac{L}{R} = \frac{60 \times 10^{-3}}{12.5 \times 10^{3}} = 4.8 \ \mu\text{sec}$$

step 8: Calculate i.

$$i = I\epsilon^{-t/\tau} = 3.9\epsilon^{-3.6/4.8}$$
$$i = 1.83 \ \text{ma}$$

step 9: Calculate e_R.

$$e_R = iR = 1.83 \times 12.5$$
$$e_R = 22.88 \ \text{v}$$

step 10: Calculate e_L.

$$e_L = -e_R = -22.88 \ \text{v}$$

problems

1. If the flux linking a coil of 3500 turns changes at a rate of 4×10^{-4} weber per second, what is the induced emf across the coil?

2. Find the inductance L in henries in the given inductor.

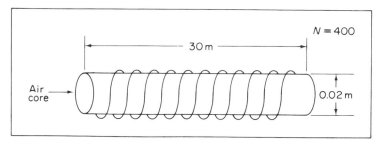

3. A coil of 640 turns has an inductance of 240 mH. If 220 turns are removed from the coil, determine the new value of inductance.

4. An air core coil of 800 turns has an inductance of 320 mH. A core material is inserted to make the relative permeability ten times the previous permeability. How many turns must be removed to obtain an inductance of 400 mH?

5. Find the voltage across a coil of 400 mH if the rate of change of current is
   (a) 25 ma/sec.
   (b) 40 ma/msec.
   (c) 200 μa/sec.

6. Find the total inductance of the following circuits.

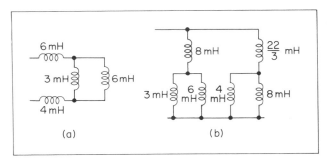

7. For the following circuit, find, after the S switch is in position 1 and after 10 μsec,

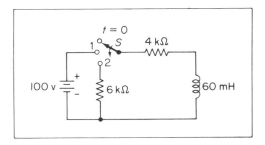

   (a) The instantaneous current.
   (b) The voltage across R.
   (c) The voltage across L.
   At 12 μsec, switch S is moved to position 2. Find, after 2 μsec,
   (a) The instantaneous current.
   (b) The voltage across R.
   (c) The voltage across L.

8. For the following circuit, determine the following after switch S is in position 1 for 240 μsec.
(**a**) The instantaneous current.
(**b**) The voltage across R.
(**c**) The voltage across L.
Move switch S to position 2. After 100 μsec, determine
(**a**) The instantaneous current.
(**b**) The voltage across R.
(**c**) The voltage across L.

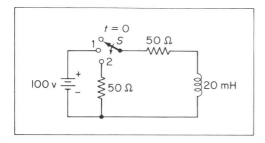

9. The voltage applied to a series RL circuit is 400 volts.
$R = 100 \, \Omega,$ $L = 400 \, \text{mH}$
Find the time required to rise to
(**a**) 40 percent of final value.
(**b**) 80 percent of final value.

10. The voltage applied to a series RL circuit is 200 volts.
$R = 40 \, \Omega,$ $L = 1 \, \text{H}$
Find the time necessary to rise to
(**a**) 25 percent of final value.
(**b**) 60 percent of final value.

11. The voltage applied to a series RL circuit is 100 volts.
$R = 50 \, \Omega,$ $t = 240 \, \mu\text{sec}$
Find the value of L required to make the current
(**a**) 1 ampere.
(**b**) 1.6 amperes.

12. For the following circuit, determine the following after switch S is in position 1 for 50 μseconds.
(**a**) The instantaneous current.
(**b**) The voltage across R.
(**c**) The voltage across L.

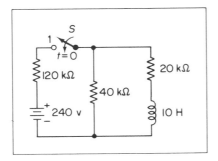

13. Switch *S* has been closed for 1 second. At this time, the switch is opened. Determine, after 80 μseconds,

(a) The instantaneous current.

(b) The voltage across *R*.

(c) The voltage across *L*.

14. For the following circuit, determine the following after switch *S* is in position 1 after 60 μseconds.

(a) The instantaneous current.

(b) The voltage across *R*.

(c) The voltage across *L*.

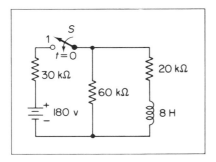

Switch *S* has been closed for 200 μseconds. It is now opened. Determine, after 40 μseconds,

(a) The instantaneous current.

(b) The voltage across *R*.

(c) The voltage across *L*.

alternating current

rotating generator

A knowledge of sinusoidal waveforms is a requirement in basic electrical technology or electrical engineering study. Heretofore, the only practical source emf that has been considered is the battery emf. Sources that generate sinusoidal waveforms utilize the principle of electromagnetic induction. The motion of the electric conductor across magnetic lines of force induces an emf in the conductor.

The basic arrangement of a simple ac generator is shown in Fig. 11–1. The loop conductor rotates in the magnetic field produced by an electromagnet. The rotating loop conductor is called the *armature* of the generator.

The rotation of the loop is counterclockwise in a circular direction. As the loop rotates through a semicircle or 180° starting from zero, the amount of voltage induced varies from zero to maximum and then back to zero. The rotation of the loop in a semicircle from 180° to 360° reverses the polarity of the induced voltage, which varies from zero to a negative maximum back to zero.

The complete revolution of the loop in a circle (0°–360°) is called a cycle. The voltage induced in the armature has an alternating waveform, as shown in Fig. 11–2. The fixed portion of the generator is called the *stator*.

definitions and nomenclature

The usual engineering expression for a sinusoidal function is

$$e = E_m \sin (\omega t \pm \varphi)$$

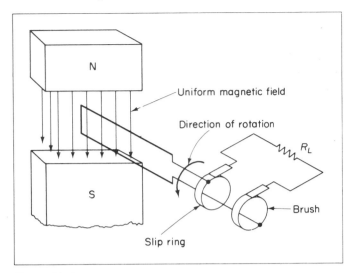

figure 11–1
ac generator

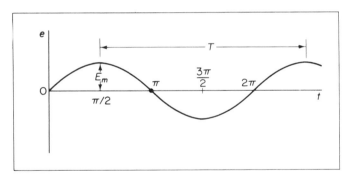

figure 11–2
generated sine wave

Using the waveform of Fig. 11–2 as a model, we have the following definitions and nomenclature.

(a) instantaneous value   the magnitude of the waveform at any instant of time. Denoted by lowercase letters (i, e, etc.).

(b)   peak value   the magnitude of the maximum value of a waveform. Denoted by uppercase letters and subscript m (E_m, I_m, etc.).

(c)   period   the time interval between successive peaks of a periodic waveform. Usually denoted by uppercase letter T.

(d) cycle that portion of a waveform contained in one period of time.

(e) frequency the number of cycles that occur in one second. For example, the power line frequency is 60 cycles per second. A modern-day abbreviation for cycles per second is the hertz, or Hz. Frequency can also be defined as the reciprocal of the period. Thus,

$$f = \frac{1}{T\,(\text{sec})}\ \text{Hz} \qquad (11\text{--}1)$$

For a 60-Hz sine waveform, the period is

$$T = \tfrac{1}{60}\ \text{sec}$$

the sine wave

The definitions and nomenclature given in the previous section may be used to describe any periodic waveform whether continuous or discontinuous. A sinusoid is a particularly useful waveform because of the simplicity of the mathematics involved. For example, the sum of two sinusoids of the same frequency will be a sinusoid. Multiplication of two sinusoids yields a sinusoid. The response of any linear, passive, bilateral network to a sinusoidal input function is also a sinusoid.

A rotating loop conductor cutting across magnetic lines of force making one complete revolution has completed one cycle of a sinusoid. In Fig. 11–3 the armature is shown in its position for one complete cycle. The generation of one complete revolution or sine wave is also shown.

The unit of measurement for the horizontal axis or abscissa is the degree or radian. The relationship between degrees and radians is based on the following formula.

$$2\pi\ (\text{radians}) = 360°$$

$$1\ \text{radian} = 57.3°$$

The angular velocity of any rotating radius vector is equal to

$$(\text{Angular velocity})\ \omega = \frac{2\pi\ (\text{radian distance})}{T\ (\text{time in seconds})}$$

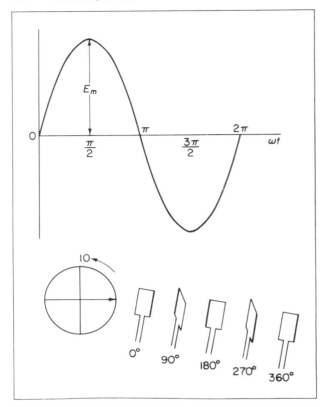

figure 11–3
a complete revolution of a loop conductor

Since $T = 1/f$, then substituting into the formula for angular velocity yields

$$\omega = 2\pi f \text{ radians/second} \tag{11-2}$$

sample problem

Determine the angular velocity of a sine wave with a frequency of 60 Hz.

solution

$$\omega = 2\pi f$$
$$\omega = 2(3.14) \times 60$$
$$\omega = 377 \text{ radians/second.}$$

sample problem

Determine the frequency of a sine wave having an angular velocity of 1256 rad/sec.

solution

$$\omega = 2\pi f$$

$$f = \frac{\omega}{2\pi} = \frac{1256}{6.28}$$

$$f = 200 \text{ Hz}$$

phase angle

The voltage waveform in Fig. 11–2 is called a sine wave, sinusoidal waveform, or sinusoid because the voltage at any instant of time is a direct function of the sine of the angle of rotation. Thus, for electrical quantities such as current and voltage, the sinusoidal equations are

$$e = E_m \sin (\omega t \pm \varphi)$$

and

$$i = I_m \sin (\omega t \pm \varphi)$$

where the capital letter with the subscript m represents maximum amplitude and the lowercase letters represent instantaneous values of current and voltage, respectively, at any instant of time t. At this point, the term alternating current will be applied only to sinusoidal ac terms. For simplicity and proper use of field terminology, the term "ac" will be used throughout, and the term "sinusoidal" will be implied.

If the phase shift is assumed zero and a value of unity is assigned to the maximum voltage amplitude, then Table 11–1 lists the numerical values for the sine of some important angles.

Suppose the armature of the generator started rotating at some voltage output rather than at zero output. As an example, assume that two ac waveforms are given by the equations

$$e_1 = E_m \sin \omega t$$

and

$$e_2 = E_m \sin (\omega t - \varphi)$$

table 11–1   sine wave values

ωt	$\sin \omega t$	ωt	$\sin \omega t$
0	0	210°	– 0.5
30°	0.5	240°	– 0.866
60°	0.866	270°	– 1.000
90°	1.000	300°	– 0.866
120°	0.866	330°	– 0.5
150°	0.5	360°	0
180°	0		

A graph of these two equations is shown in Fig. 11–4. The phase difference between the two waveforms is defined by the angle φ.

The phase of a periodic waveform is the fractional part of a period that is displaced on either side of zero. The initial phase is defined at $\omega t = 0$; that is, at $\omega t = 0$, $e_1 = 0$, and $e_2 = -E_m \sin \varphi$.

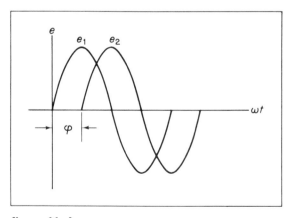

figure 11–4
phase relationships

The terms *lead* and *lag* are used to define phase relationships between two sinusoidal waveforms of the same frequency. Consider the equations specifying the phase relationships between two sinusoidal quantities of the same frequency, as shown in Fig. 11–5.
Thus,

$$e_1 = E_m \sin (\omega t + \varphi_1)$$

$$e_2 = E_m \sin (\omega t - \varphi_2)$$

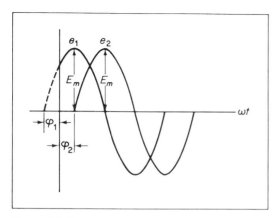

figure 11–5
phase relationship between two voltages

The phase difference between two sinusoidal quantities is the difference between their phases at any instant.
Thus,

$$(\omega t + \varphi_1) - (\omega t - \varphi_2) = \varphi_1 + \varphi_2$$

In this case, voltage e_1 leads voltage e_2 by the phase angle $(\varphi_1 + \varphi_2)$. On the other hand, voltage e_2 lags voltage e_1 by the phase angle $(\varphi_1 + \varphi_2)$. An illustrative problem will demonstrate the theory.

sample problem

Two waveforms are shown. What are the equations for each voltage? What is their phase difference?

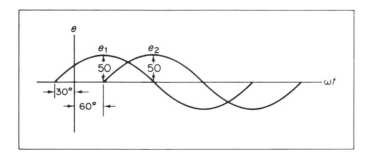

solution

step 1 :   From the diagram given,

$$e_1 = 50 \sin (\omega t + 30°)$$
$$e_2 = 50 \sin (\omega t - 60°)$$

step 2 :   The phase difference is

$$(\omega t + 30°) - (\omega t - 60°) = 90°$$

Note that e_1 leads e_2 by 90°, or e_2 lags e_1 by 90°.

---

sample problem

Given the diagram shown, determine: (a) the voltage and current equations; (b) the phase difference.

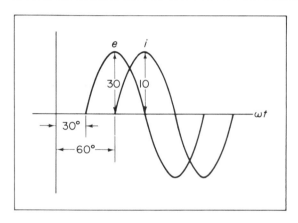

solution

step 1 : $e = 30 \sin (\omega t - 30°)$
$i = 10 \sin (\omega t - 60°)$

step 2 :   Voltage e leads i by 30°; current i lags e by 30°.

alternating currents

Since an alternating sine wave can have a multitude of instantaneous values through a cycle, the peak magnitude may be used as a method of comparing two ac waveforms. The peak value of an ac waveform is the maximum value of the wave, as shown in Fig. 11–6.

The peak value of a sine wave applies to either the negative or the positive peak, since they are equal in value. Because of this equality, the peak-to-peak value of a sine wave is twice the peak value.

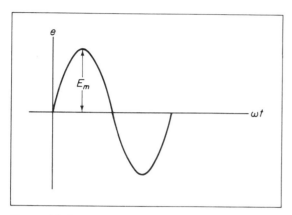

figure 11–6
ac waveform

During the first half or positive cycle of voltage, the direction of conventional current through a resistance or resistive element is positive, or follows the voltage. Consequently, the voltage and current are said to be in phase across a resistive element. The amplitude or peak magnitude of the current is equal to

$$I_m = \frac{E_m}{R}$$

On the next half cycle or negative portion of the waveform, the polarity of the voltage reverses. With reversed polarity, the current direction reverses accordingly. The amplitude of the current in the reverse direction is also equal to

$$I_m = \frac{E_m}{R}$$

The voltage and current waveforms are shown in Fig. 11–7.

It is evident that both half cycles of voltage are equally effective in producing current. Although the amplitudes are equal, the current polarities are direct functions of the applied voltage polarity. As the ac wave varies from instant to instant, some method must be established of determining the peak, effective, and average values of the ac waveform.

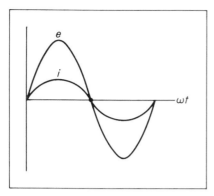

figure 11–7
current and voltage in phase across a resistor

The heating effect of an ac wave's being independent of current polarity can be utilized as a criterion for both ac and dc waveforms. Consequently, an ac current is said to have an effective value of 1 ampere when it produces a heating effect across a specified resistance at the same average rate as 1 ampere of dc. Thus, consider the circuit shown in Fig. 11–8.

Close switch 1. A current flows through the lamp load designated by R_L, and the lamp is lit. The dc current in the circuit is equal to

$$I_{dc} = \frac{E}{R_L}$$

Open switch 1, and the lamp is out. Close switch 2 and vary the generator voltage e until the lamp brilliance has the same intensity as it did previously with the dc circuit. When this is accomplished, the average electrical power

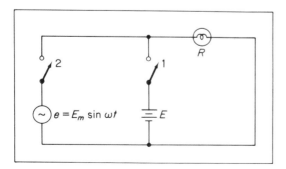

figure 11–8
demonstration of effective current

dissipated across the lamp by the ac source is exactly equal to the dc source power.

Since the heating effect of a dc source is proportional to the square of the current or $I_{dc}^2 R_L$, the heating effect of an ac source at any instant is proportional to the square of the current at that instant. The effective value of an ac wave must be considered on the basis of a square value. Thus, investigating the two waveforms shown in Fig. 11–9, note that waveform 2 is the square of waveform 1. The power delivered by the ac source is

$$P_{ac} = (I_m \sin \omega t)^2 \, R$$

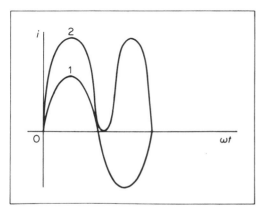

figure 11–9
ac wave and its square

Using the trigonometric identity, we find that

$$\sin^2 \omega t = \tfrac{1}{2} (1 - \cos 2\omega t)$$

Then

$$P_{ac} = \frac{I_m^2 R}{2} (1 - \cos 2\omega t)$$

It is obvious that the average power in a resistor through which an ac current is flowing is equal to one-half the peak value of power. If we average a sine or cosine function over one complete cycle, the value must be zero, since the area beneath the positive cycle must be exactly equal to the area beneath the negative half cycle. Thus, the result is

$$P_{ac} = \frac{I_m^2 R}{2} \text{ watts}$$

In a dc and ac circuit, the power delivered to a resistive load is equated by

$$I_{dc}^2 R = \frac{I_m^2}{2} R$$

The equivalent dc or the effective value of the ac current is

$$I_{eff} = I_{dc} = \frac{I_m}{\sqrt{2}} = 0.707 \, I_m \qquad (11\text{–}3)$$

The voltage relationships are

$$E_{eff} = 0.707 \, E_m$$

The effective value of a sinusoidal or nonsinusoidal waveform that is periodic can be evaluated by

$$I_{eff} = \sqrt{\frac{1}{T} \int_0^T [i(t)]^2 \, dt} \qquad (11\text{–}4)$$

or

$$I_{eff} = \sqrt{\frac{1}{2\pi} \int_0^{2\pi} [i(\omega t)]^2 \, d(\omega t)} \qquad (11\text{–}5)$$

The procedure used in evaluating the effective value of any wave is

step 1

Square the ordinates or $i(\omega t)$ values.

step 2

Find the area under the $[i(\omega t)]^2$ curve.

step 3

Divide the total area by the total period for one cycle.

step 4

Take the square root of the mean value.

The name "root mean square" or rms value is derived from this procedure and is usually used in place of effective value.

---

sample problem

Find the rms value of the given waveform.

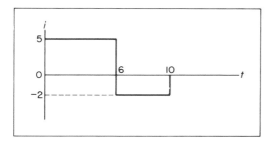

solution

step 1: Determine $[i(t)]^2$.

$$[i_1(t)]^2 = 25$$
$$[i_2(t)]^2 = 4$$

step 2: Determine the area of the waveform.

$$A_1 = [i_1(t)]^2 6 = 25 \times 6 = 150$$
$$A_2 = [i_2(t)]^2 4 = 4 \times 4 = 16$$

step 3: Divide the total area by the period.

$$\frac{A_1 + A_2}{T} = \frac{150 + 16}{10} = 16.6$$

step 4: Determine the rms value of the waveform.

$$I_{\text{rms}} = \sqrt{16.6} = 4.08 \text{ A}$$

---

sample problem

Find the rms value for the voltage wave shown.

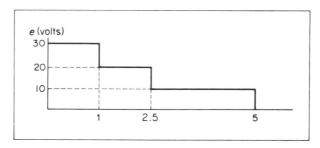

solution

step 1: Determine $[e(t)]^2$.

$$[e_1(t)]^2 = (30)^2 = 900$$
$$[e_2(t)]^2 = (20)^2 = 400$$
$$[e_3(t)]^2 = (10)^2 = 100$$

step 2: Determine the total area.

$$A_1 = [e_1(t)]^2 \, 1 = 900$$
$$A_2 = [e_2(t)]^2 \, 1.5 = 600$$
$$A_3 = [e_3(t)]^2 \, 2.5 = 250$$

step 3: Divide the total area by the period.

$$\frac{A_1 + A_2 + A_3}{T} = \frac{900 + 600 + 250}{5} = 350$$

step 4: Calculate E_{rms}.

$$E_{rms} = \sqrt{350} = 18.75 \text{ v}$$

average value

The effective value of a waveform was determined by calculating the average power over one cycle and dividing by the period. An average value is defined as the arithmetic sum of all the values divided by the total number of values used in determining the sum. The average value of a sine wave over one complete cycle must be zero, since the area under the positive half cycle is equal to the area of the negative half cycle.

The average value of one half cycle in which the period of the path is π radians (if the presence of the negative half cycle is ignored) is

$$E_{av} = \frac{2}{\pi} E_m = 0.636 \, E_m \qquad (11\text{–}6)$$

If the length of path is considered to be 2π, then the average value for one half cycle is

$$E_{av} = \frac{1}{\pi} E_m = 0.318 E_m$$

The average value of a waveform can be determined graphically by finding the area beneath the curve and dividing the total area by the period.

It should be noted that this topic was included for completeness of subject coverage and that the use of the average value of a waveform is extremely limited.

power and power factor

In the section on dc power relationships, the power delivered to a resistive load was equal to the product of the voltage across the load and the current flowing through the load. Thus,

$$P = EI \text{ w}$$

This rule is still applicable to ac circuits, provided that we consider the instantaneous values of voltage and current. Thus,

$$p = ei \text{ w}$$

The mathematical relationship for power in a network can be determined as follows. Let $e = E_m \sin \alpha$ be the applied voltage and $i = I_m \sin (\alpha + \varphi)$ be the current flowing in the circuit. Then

$$p = (E_m \sin \alpha) \, I_m \sin (\alpha + \varphi)$$

Using the trigonometric identity, we have

$$\sin A \sin B = \tfrac{1}{2} [\cos (A - B) - \cos (A + B)]$$

where

$$A = \alpha \quad \text{and} \quad B = \alpha + \varphi$$

Then

$$p = \frac{E_m \, I_m}{2} [\cos \varphi - \cos (2\alpha + \varphi)]$$

or

$$p = \frac{E_m \, I_m}{\sqrt{2} \, (\sqrt{2})} [\cos \varphi - \cos (2\alpha + \varphi)]$$

Note that

$$E_{rms} = \frac{E_m}{\sqrt{2}} \quad \text{and} \quad I_{rms} = \frac{I_m}{\sqrt{2}}$$

The angle φ is the phase angle difference between the voltage and current. If the phase difference is zero, then the power curve shown in Fig. 11–10 is sinusoidal in shape and has double the frequency as compared to the voltage or current sinusoid.

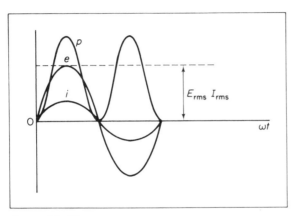

figure 11–10
power relationships in a resistor

The equation for power under this condition is

$$p = E_{rms} I_{rms} (1 - \cos 2\alpha)$$

The average value of power dissipated across a resistor is determined by

$$P_{av} = E_{rms} I_{rms} \text{ w} \tag{11–7}$$

If the angle φ is unequal to zero (the phase angle existent between voltage and current is not zero), the general equation for total power in a circuit is

$$P = E_{rms} I_{rms} [\cos \varphi - \cos (2\alpha + \varphi)]$$

The average power is then equal to

$$P_{av} = E_{rms} I_{rms} \cos \varphi \text{ w} \tag{11–8}$$

The quantity $\cos \varphi$ is called the power factor (abbreviated P.F.) of the circuit.

$$\cos \varphi = \text{P.F.} = \frac{P_{av}}{E_{rms} \, I_{rms}}$$

The power factor of a circuit can never exceed unity but is usually expressed as a decimal or percentage.

For the students who can handle calculus, the general equation for the (average) power is

$$P_{av} = \frac{1}{T} \int_0^t p \, dt = \frac{1}{T} \int_0^t ei \, dt$$

or

$$P_{av} = \frac{1}{2\pi} \int_0^{2\pi} p \, d(\omega t)$$

ac circuits with resistance

Assume that a sinusoidal current supplied by an ac generator is flowing through a resistor as shown in Fig. 11–11. The resultant waveforms both are sinusoidal, having the same frequency with zero phase difference. Using Ohm's law, we say that the relationship that exists is

$$e = iR$$

and if $i = I_m \sin \omega t$, then

$$e = I_m R \sin \omega t$$

or

$$e = E_m \sin \omega t$$

where

$$E_m = I_m R.$$

Since R is defined as the ratio of voltage to current, then

$$\boxed{R = \frac{e}{i} = \frac{E_m}{I_m} = \frac{E_{rms}}{I_{rms}} \, \Omega}$$

(11–9)

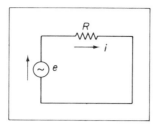

figure 11–11
ac circuit with resistance

The average value of power is

$$P_{av} = E_{rms}I_{rms} = I_{rms}^2 R = \frac{E_{rms}^2}{R} \text{ w}$$ (11–10)

The waveforms of voltage, current, and power are shown in Fig. 11–12.

ac circuit with inductance

In a purely inductive circuit, the applied voltage is equal and opposite to the self-induced voltage according to Lenz's law. The relationship among voltage, current, and inductance is defined by

$$L\frac{di}{dt} = e$$

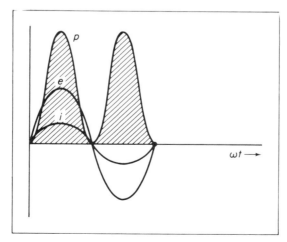

figure 11–12
voltage, current, and power relationship across *R*

Consider the pure L circuit with an ac source of energy as input as shown in Fig. 11–13.

If the current flowing in a pure L circuit is sinusoidal and defined by

$$i = I_m \sin \omega t$$

then the voltage induced in the coil is equal to

$$e = \omega L I m \cos \omega t$$

or

$$\boxed{e = \omega L I_m \sin (\omega t + 90°)} \tag{11–11}$$

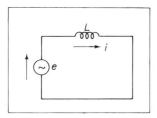

figure 11–13
ac applied to a pure L circuit

Note that the applied emf leads the current by 90° through an inductance. From these equations, it is evident that for sinusoidal waveforms

$$e = E_m \sin (\omega t + 90°)$$

where $E_m = \omega L I_m$. The waveforms of voltage, current, and power are shown in Fig. 11–14.

The quantity ωL is called the *reactance* of the inductor and is represented by the letter symbol X_L. The units of X_L must be in ohms. Since the phase difference between voltage and current is 90°, the power factor is zero, and the average power dissipated across an inductor is zero.

It is the opposition to the flow of current that results in energy's being stored and returned to the circuit because of the variation of the magnetic field of the inductor. This opposition cannot be resistive, since voltage and current are in phase across a resistor, whereas the current lags the voltage by 90° across an inductance. The basic equations for the voltage and current through an inductance are

$$e = E_m \sin \omega t$$

$$i = I_m \sin (\omega t - 90°)$$

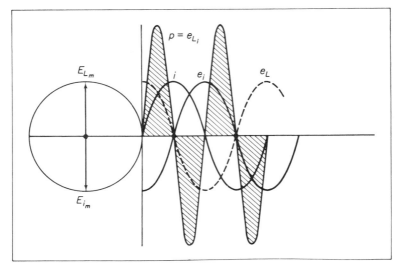

figure 11–14
voltage, current, and power across an inductor

and

$$E_m = I_m \omega L$$

The instantaneous power is

$$p = ei = E_m I_m \sin \omega t \sin (\omega t - 90°)$$

$$p = \frac{E_m I_m}{2} [\cos 90° - \cos (2\omega t - 90°)]$$

$$p = \frac{E_m I_m}{2} (0 + \sin 2\omega t)$$

The instantaneous power passes through two complete cycles for each cycle of voltage or current. The average power is

$$P_{av} = \frac{1}{2\pi} \int_0^{2\pi} p \, d(\omega t) = 0$$

ac applied to a pure C circuit

Connect a pure capacitance across an ac source, as shown in Fig. 11–15. The opposing emf of a capacitor is given by the equation

$$e_c = \frac{q}{C}$$

Kirchhoff's voltage law requires that the voltage across the capacitor must be exactly equal to the applied voltage at any instant of time. To meet this requirement, the capacitor must charge and discharge at a rate that follows the applied potential at every instant. Therefore, the time rate variation of charge is

$$\frac{dq}{dt} = C\frac{de_c}{dt}$$

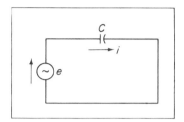

figure 11–15
ac applied to a pure C circuit

Since

$$i = \frac{dq}{dt},$$

then

$$i = C\frac{de_c}{dt}$$

If $e = E_m \sin \omega t$, the value of i is

$$i = \omega C E_m \sin (\omega t + 90°)$$

The waveforms of current, voltage, and power are shown in Fig. 11–16. It is evident from the given equations that the current flowing through a pure capacitance leads the voltage across the capacitor by 90°.

When we connect a pure capacitance across an ac source, the current is a sinusoid leading the voltage wave by 90°. The ratio of voltage to current represents the opposition of the capacitor to the flow of current. Therefore,

$$\frac{E_m}{I_m} = \frac{1}{\omega C} = X_c \qquad (11\text{–}12)$$

The quantity $1/\omega C$ is called the *capacitive reactance* and is represented by the letter symbol X_c. Like the inductor, the capacitor does not dissipate power but stores it in the dielectric field for charge and discharge operations.

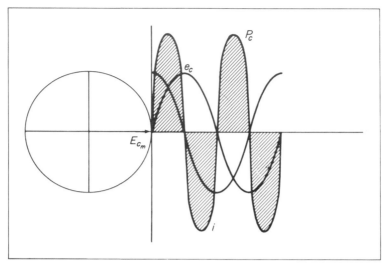

figure 11–16
voltage, current, and power across a capacitor

Again, since the phase difference between voltage and current is 90°, then the power factor is zero and the average power dissipated in the capacitor is zero. The basic equations for the current and voltage across a capacitor are

$$e = E_m \sin \omega t$$

$$i = I_m \sin (\omega t + 90°)$$

$$E_m = \frac{I_m}{\omega C} = I_m X_c$$

The instantaneous power is

$$p = ei = (E_m \sin \omega t)I_m \sin (\omega t + 90°)$$

$$p = \frac{E_m I_m}{2} [\cos 90° - \cos (2\omega t + 90°)]$$

$$p = E_{rms} I_{rms}(0 + \sin 2\omega t)$$

The instantaneous power passes through two complete cycles for each cycle of current or voltage. The average power is

$$P_{av} = \frac{1}{2\pi} \int_0^{2\pi} p \, d(\omega t) = 0$$

An illustrative problem will demonstrate the theory.

A voltage of $e = 100 \sin 754t$ is applied. Determine the instantaneous and peak values of current when (a) $R = 50\ \Omega$, (b) $L = 1.59$ h, and (c) $C = 1.59\ \mu\text{f}$.

solution

step 1: Calculate i through R.

$$i = \frac{e}{R} = \frac{100}{50} \sin 754t$$

step 2: Calculate I_m through R.

$$I_m = \frac{E_m}{R} = \frac{100}{50} = 2\ \text{A}$$

step 3: Calculate i and I_m through L.

$$X_L = \omega L = 2\pi f L = 754 \times 1.59 = 1200\ \Omega$$

$$i = \frac{E_m}{X_L} \sin (754t - 90°)$$

Since the current lags the voltage through an inductor,

$$i = 0.0833 \sin (754t - 90°)\ \text{A}$$

and

$$I_m = 0.0833\ \text{A}$$

step 4: Calculate i and I_m through C.

$$X_c = \frac{1}{\omega C} = 833\ \Omega$$

$$i = \frac{E_m}{X_c} \sin (754t + 90°)$$

Since the current leads the voltage through a capacitor,

$$i = 0.12 \sin (754t + 90°)$$

and

$$I_m = 0.12\ \text{A}$$

problems

1. Determine the period of waveforms having a frequency of:
   (a) 20 Hz   (d) 1 MHz
   (b) 10 kHz   (e) 25 MHz
   (c) 400 kHz

2. Determine the frequency of the waveforms having a period of:
   (a) 50 msec   (d) 1 μsec
   (b) 10 μsec   (e) 40 nanosec
   (c) 25 μsec

3. The angular velocity of a waveform is 800π rad/sec.
   Determine: (a) the frequency and (b) the period.

4. The angular velocity of a waveform is 2,512,000 rad/sec.
   Determine: (a) the frequency and (b) the period.

5. What is the frequency of the following waveforms?
   (a) $i = 10 \sin (377t + 10°)$
   (b) $e = 20 \sin 1590t$
   (c) $i = 30 \sin (1508t - 20°)$
   (d) $e = 40 \sin 6032t$

6. Two currents have the respective equations

$$i_1 = 75 \sin \left(200\pi t + \frac{\pi}{6} \right)$$

$$i_2 = 30 \sin \left(200\pi t - \frac{\pi}{4} \right)$$

   (a) Determine the phase difference between the currents.
   (b) Does current i_1 lead or lag i_2?

7. The equation for an ac current is

$$i = 30 \sin \left(100t - \frac{\pi}{6} \right) \text{A}$$

   Determine the value of current at t equal to:
   (a) 0.02 sec   (b) 0.06 sec
   (c) $\dfrac{2\pi}{300}$ sec   (d) $\dfrac{2\pi}{600}$ sec

8. Determine the effective value of the waveform shown.

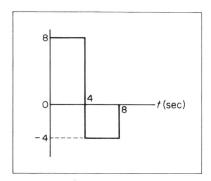

9. Determine the effective value of the waveform shown.

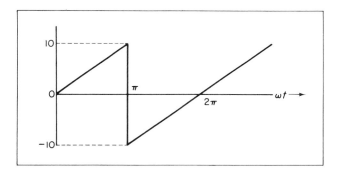

10. Determine the effective value of the waveform shown.

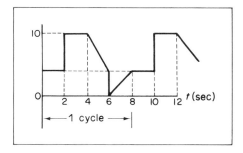

11. Determine the effective value of the waveform shown.

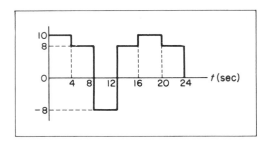

12. Determine the effective value of the waveform shown.

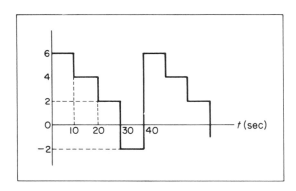

13. Write the sinusoidal equations for voltage having the following effective values at a frequency of 120 Hz and having zero phase shift.

   (**a**) 141.4 v   (**c**) 424.2 v
   (**b**) 70.7 v   (**d**) 353.5 mv

14. Find the average power loss in watts for the following:

   (**a**) $e = 40 \sin (\omega t + 30°)$
        $i = 25 \sin (\omega t - 25°)$
   (**b**) $e = 50 \sin (\omega t - 35°)$
        $i = 20 \sin (\omega t - 15°)$
   (**c**) $e = 30 \sin (\omega t + 65°)$
        $i = 15 \sin (\omega t - 10°)$
   (**d**) $e = 0.05 \sin (\omega t + 15°)$
        $i = 0.003 \sin (\omega t - 22°)$

15. A circuit dissipates 100 watts of average power. The effective voltage applied is 220 volts, and the current is 1 A rms. Determine the power factor.

16. The power factor of a circuit is 0.866. The power delivered to the load is 500 watts average. If the input voltage is 100 sin $(\omega t + \pi/3)$, determine the sinusoidal equation for current.

17. The current and voltage equations are
$$e = 44 \sin (\omega t + 35°)$$
$$i = 0.0056 \sin (\omega t + 35°)$$
Determine the average output power.

18. The power factor of a circuit is 0.7 and the average power delivered to the load is 200 watts. If the input voltage has an effective value of 100 volts, find the sinusoidal expression for the current.

19. A sine wave of voltage applied to a linear resistance of 25 ohms has the form $e = 200 \sin (160\pi t - 30°)$ volts. Calculate
   (a) The instantaneous current
   (b) The average power
   (c) The power at $t = 3.125$ msec
   (d) The frequency

20. A capacitance of 7.4 μf has a voltage applied equal to $e = 430 \cos 1000\pi t$. Calculate
   (a) The instantaneous current
   (b) The capacitive reactance
   (c) The power at $t = 250$ μsec

21. A capacitance of 1400 pf has a voltage applied equal to $e = 250 \sin 2\pi \times 10^6 t$. Calculate
   (a) The instantaneous current
   (b) The capacitive reactance
   (c) The power at $t = 33.3$ nanosec

22. Determine the frequency at which the reactance of a 1.2-μf capacitor is 4000 ohms.

23. The reactance of a capacitance is 5000 ohms at a frequency of 30 kHz. Determine the capacitor used.

24. A coil of 63 μH has a voltage applied equal to $e = 380 \sin 2\pi \times 4 \times 10^6 t$. Calculate
   (a) The instantaneous current
   (b) The inductive reactance
   (c) The power at $t = 41.8$ nanosec

25. A coil of 84 μH has a voltage applied equal to $e = 200 \sin 2\pi \times 4.55 \times 10^6 t$. Calculate
   (a) The inductive reactance
   (b) The instantaneous current
   (c) The average power

26. Determine the frequency at which the reactance of a 35 mH coil is 25 kΩ.

27. The reactance of a coil is 10,000 ohms at a frequency of 500 Hz. Determine the inductor size or value.

12

phasor algebra

complex quantities

Alternating currents and voltages are phasor quantities rather than scalar and must be treated by special mathematical techniques. The justification for the use of complex numbers is that it provides a simplified method to the solution of ac circuits.

A complex number is a point that is represented by two components, one along and the other at right angles to some arbitrary axis of reference. This point can also determine a radius drawn from the origin of the arbitrary axis to the point. The horizontal axis or the *abscissa* contains all real numbers, whereas the vertical axis or the *ordinate* is called the imaginary axis, since it contains so-called imaginary numbers.

In the complex plane shown in Fig. 12–1, all real and imaginary numbers in the first quadrant are positive; the second quadrant contains a negative real number and a positive imaginary number. The third quadrant contains negative real and imaginary numbers, and in the fourth quadrant the real numbers are positive and the imaginary numbers are negative.

In electrical engineering, the symbol j in front of the R or X component denotes a vertical element or imaginary number. Consider the complex plane shown in Fig. 12–2. In this complex plane exists a point whose coordinates are three horizontal units and one vertical unit. This point represents a radius phasor equal to $3+j1$. Since j is an imaginary number, then the definition of j is

$$j = \sqrt{-1}$$

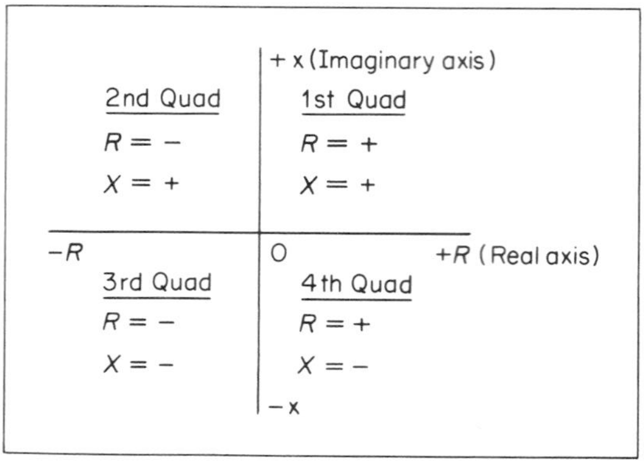

	+ x (Imaginary axis)
2nd Quad R = − X = +	1st Quad R = + X = +
3rd Quad R = − X = −	4th Quad R = + X = −

−R   O   +R (Real axis)

− x

figure 12–1
complex plane

and

$$j^2 = -1$$

$$j^3 = -j = -\sqrt{-1}$$

$$j^4 = +1$$

It is evident that when a quantity is multiplied by j, the quantity is rotated through an angle of 90° in a counterclockwise direction.

A complex quantity is a number consisting of a real part R and an imaginary part or the j portion, represented by jX.

A phasor may be represented by any one of the following forms: (a) rectangular, (b) polar, (c) trigonometric, (d) exponential.

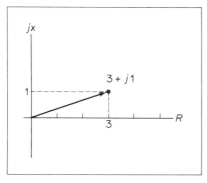

figure 12–2
use of coordinates in a complex plane

rectangular form

The phasor $Z = 3+j4$ as shown in Fig. 12–3 is said to be expressed in rectangular form. This form can be used to describe a radius phasor in any one of the four quadrants. Thus,

	R	X
First quadrant	$+3$	$+j4$
Second quadrant	-3	$+j4$
Third quadrant	-3	$-j4$
Fourth quadrant	$+3$	$-j4$

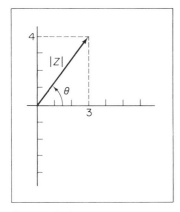

figure 12–3
rectangular form of complex number

trigonometric form

By inspection of Fig. 12–3, the components of the phasor Z can be expressed in terms of the magnitude of Z, symbolized by $|Z|$. Thus,

$$R = |Z| \cos \theta$$

$$jX = j|Z| \sin \theta$$

where

$$Z = \sqrt{R^2 + X^2} \quad \text{(magnitude)}$$

$$\theta = \tan^{-1} \frac{X}{R} \quad \text{(direction angle)}$$

Consequently, $Z = R+jX = Z (\cos \theta + j \sin \theta)$. This form is called the trigonometric form and is also used to represent a radius phasor. The position of the magnitude of Z is determined by the direction angle θ. Rotating the radius phasor through a negative or clockwise direction results in

$$Z = |Z| [\cos(-\theta) + j \sin(-\theta)]$$

or

$$Z = |Z| (\cos\theta - j \sin\theta)$$

A graph of the radius phasors is shown in Fig. 12–4.

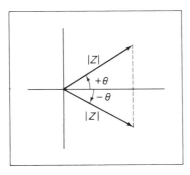

figure 12–4
magnitude and direction of a radius phasor

polar form

A convenient method of writing a radius phasor quantity known as the polar form is

$$Z = |Z| \,\underline{/\theta°}$$

This notation should be interpreted as the magnitude of Z making an angle of θ degrees with the horizontal axis in the counterclockwise direction of rotation. This notation does not imply a product function and cannot be treated as such. For example, the complex quantity $Z = 3 + j4$ can be written in polar form as

$$Z = 5\,\underline{/53.1°}$$

where

$$|Z| = \sqrt{3^2 + 4^2} = 5$$
$$\theta = \tan^{-1}\tfrac{4}{3} = 53.1°$$

This phasor has a magnitude of 5 and makes an angle of $53.1°$ with the horizontal axis.

The exponential form for writing a complex quantity is

$$Z = |Z| \epsilon^{j\theta}$$

This relationship is derived by expanding the $\cos \theta$ and $\sin \theta$ functions into an infinite series by means of calculus. The next step is to expand ϵ^θ into an infinite series and apply McLaurin's theorem. The final result is shown as

$$Z = |Z| \epsilon^{j\theta}$$

conversion methods

It is evident that a complex quantity can be represented in any one of the complex forms. Consequently, a method of converting from one form to another must exist. Thus,

$$Z = \quad R+jX \quad = |Z|\underline{/\theta} = |Z|(\cos \theta + j \sin \theta) = \quad |Z| \epsilon^{j\theta}$$

| Rectangular form | Polar form | Trigonometric form | Exponential form |

The slide rule is an invaluable tool for conversion operations. For proper usage of the slide rule, see the appendix. To convert from rectangular form to polar form, the following procedure is utilized. Thus,

$$Z = R+jX = |Z| \underline{/\theta}$$

To determine the magnitude of Z, the following formula must be applied.

$$|Z| = \sqrt{R^2 + X^2}$$

The direction angle θ is determined by

$$\theta = \tan^{-1} \frac{X}{R}$$

To convert from the polar form to the rectangular, the phasor is set up in the trigonometric form and then evaluated in terms of real and imaginary components. Thus,

$$Z = |Z|\underline{/\theta} = R+jX$$

and

$$R = |Z| \cos \theta$$

$$X = |Z| \sin \theta$$

Some illustrative problems will demonstrate the theory.

---

sample problem

Convert from the rectangular form to the polar form the following phasors:

        (a) $4-j3$ (b) $10+j5$ (c) $Z = -2-j3$

---

solution

    (a) $Z = 4-j3$

$$|Z| = \sqrt{4^2+3^2} = 5$$

$$\theta = \tan^{-1} \tfrac{3}{4} = 36.9°$$

$$Z = 5\underline{/-36.9°}$$

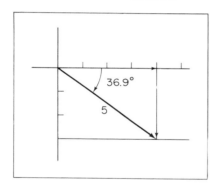

    (b) $Z = 10+j5$

$$|Z| = \sqrt{10^2+5^2} = 11.2$$

$$\theta = \tan^{-1} \tfrac{5}{10} = 26.6°$$

$$Z = 11.2\underline{/26.6°}$$

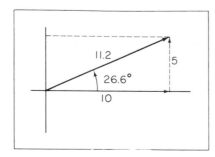

(c) $Z = -2 - j3$

$$|Z| = \sqrt{2^2 + 3^2} = 3.6$$

$$\theta = \tan^{-1} \tfrac{3}{2} = 56.4°$$

$$Z = 3.6\underline{/236.4°}$$

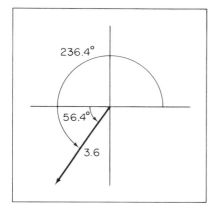

sample problem

Convert from polar form to rectangular form the following:

(a) $6\underline{/43°}$ (b) $10\underline{/86°}$ (3) $8\underline{/-18°}$

solution

(a) $Z = 6\underline{/43°} = 6(\cos 43° + j \sin 43°)$

$$Z = 4.39 + j4.1$$

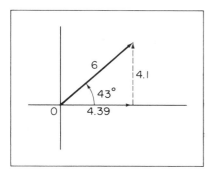

(b) $Z = 10\underline{/86°} = 10(\cos 86° + j \sin 86°)$

$Z = 0.7 + j9.93$

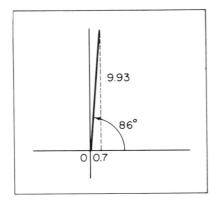

(c) $Z = 8\underline{/-18°} = 8(\cos 18° - j \sin 18°)$

$Z = 7.6 - j2.47$

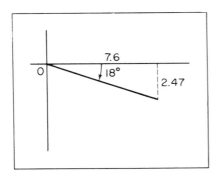

mathematical operations

addition

The addition of two complex numbers can be performed both graphically and analytically. Using the analytic technique, assume that there are two complex numbers having the following values.

$$Z_1 = R_1 + jX_1 \quad \text{and} \quad Z_2 = R_2 + jX_2$$

In the addition of complex numbers, add the real parts of the complex quantity and then add the j parts independently. The result of the addition is

$$Z_3 = Z_1 + Z_2 = (R_1 + R_2) + j(X_1 + X_2)$$

The graphical technique of addition is performed in the following manner.

step 1

Place all phasors with their respective magnitudes and direction angles on the complex plane as shown in Fig. 12–5.

step 2

Form a parallelogram as shown in Fig. 12–5. This is performed by adding phasor Z_1 to the tip of phasor Z_2.

step 3

Starting from the origin, the diagonal of the parallelogram is the resultant phasor Z_3.

step 4

It is evident from the diagram that the real part of the resultant, or Z_3, is equal to the sum of the real parts of phasors Z_1 and Z_2, respectively. The j parts of Z_1 and Z_2 when added together comprise the j part of Z_3. Consequently, the magnitude of Z_3 is

$$Z_3 = \sqrt{(R_1 + R_2)^2 + (X_1 + X_2)^2}$$

The direction angle of Z_3 is defined by

$$\theta = \tan^{-1} \frac{(X_1 + X_2)}{(R_1 + R_2)}$$

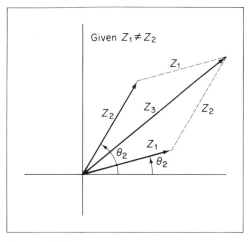

figure 12–5
phasor addition

Note that if phasors Z_1 and Z_2 are presented in either polar or exponential form, they must be converted to rectangular form before the addition can be performed. The technique of addition can be performed regardless of the number of phasors involved. All the real parts and all the j or imaginary parts of the complex numbers are summed together with all positive and negative signs accounted for.

subtraction

In a manner similar to the subtraction of algebraic quantities, one phasor may be subtracted from another by changing the sign of the phasor and adding. Thus, if $Z_1 = R_1 + jX_1$ and $Z_2 = R_2 + jX$, then

and

$$Z_3 = Z_1 - Z_2 = (R_1 - R_2) + j(X_1 - X_2)$$

$$Z_3 = |Z_3| \underline{/\theta_3}$$

$$|Z_3| = \sqrt{(R_1 - R_2)^2 + (X_1 - X_2)^2}$$

$$\theta_3 = \tan^{-1} \frac{(X_1 - X_2)}{(R_1 - R_2)}$$

multiplication

For multiplication, the complex quantities may be expressed in either the rectangular or the polar form. Each form used has its own advantages and disadvantages. Consider the product of two phasors in the rectangular form, where

$$Z_1 = R_1 + jX_1 \quad \text{and} \quad Z_2 = |Z_2| \underline{/\theta_2}$$

This product cannot be handled directly, since Z_2 must be converted to rectangular form. Using the trigonometric form to convert from polar form to rectangular form yields

$$Z_2 = |Z_2|(\cos \theta_2 + j \sin \theta_2) = R_2 + jX_2$$

Let

$$Z_3 = Z_1 \times Z_2 = |Z_3| \underline{/\theta_3}$$

and

$$R_1 + jX_1$$
$$R_2 + jX_2$$

$$\overline{R_1 R_2 + jR_2 X_1}$$
$$\quad\quad + jR_1 X_2 + j^2 X_1 X_2$$

$$\overline{R_1 R_2 + j^2 X_1 X_2 + j(R_1 X_2 + R_2 X_1)}$$

Since

$$j^2 = -1,$$

then

$$Z_3 = (R_1 R_2 - X_1 X_2) + j(R_1 X_2 + R_2 X_1) = |Z_3|\ \underline{/\theta_3}$$

where

$$Z_3 = \sqrt{(R_1 R_2 - X_1 X_2)^2 + (R_1 X_2 + R_2 X_1)^2}$$

$$\theta_3 = \tan^{-1}\left(\frac{R_1 X_2 + R_2 X_1}{R_1 R_2 - X_1 X_2}\right)$$

Some illustrative examples will demonstrate the theory.

---

sample problem

Find the product of the following phasors:

(a) $Z_1 = 3 + j4$ and $Z_2 = 5 - j2$
(b) $Z_1 = -2 - j3$ and $Z_2 = 4 + j9$
(c) $Z_1 = 2 - j7$ and $Z_2 = 3 + j5$

---

solution

(a) $Z_3 = Z_1 \times Z_2$

$$3 + j4$$
$$5 - j2$$
$$\overline{}$$
$$15 + j20$$
$$\quad -j6 - j^2 8$$
$$\overline{}$$

$$Z_3 = 23 + j14$$

(b) $Z_3 = Z_1 \times Z_2$

$$-2 - j3$$
$$4 + j9$$
$$\overline{}$$
$$-8 - j12$$
$$\quad -j18 - j^2 27$$
$$\overline{}$$

$$Z_3 = +19 - j30$$

(c) $Z_3 = Z_1 \times Z_2$

$$2 - j7$$
$$3 + j5$$
$$\overline{}$$
$$6 - j21$$
$$\quad +j10 - j^2 35$$
$$\overline{}$$

$$Z_3 = 41 - j11$$

If the phasors are given in polar form, the magnitudes are multiplied and the resultant phase angle is the algebraic sum of the original angles. Thus, if $Z_1 = |Z_1| \underline{/\theta_1}$ and $Z_2 = |Z_2| \underline{/\theta_2}$, then,

$$Z_3 = |Z_1| \, |Z_2| \, \underline{/\theta_1 + \theta_2}$$

Some illustrative problems will demonstrate the theory.

---

sample problem

Multiply the following phasors in polar form only:

   (a) $(43\underline{/38°})(16\underline{/-10°})$

   (b) $(10\underline{/-30°})(14.14 + j\,14.14)(18\underline{/-55°})$

   (c) $(55\underline{/20°})^2(15\underline{/-10°})$

---

solution

   (a) $Z_3 = (43\underline{/38°})(16\underline{/-10°})$

   $\quad Z_3 = 688\underline{/28°}$

   (b) $Z_3 = (10\underline{/-30°})(20\underline{/45°})(18\underline{/-55°}$

   $\quad Z_3 = 3600\underline{/-40°}$

   (c) $Z_3 = (55\underline{/20°})^2(15\underline{/-10°})$

   $\quad Z_3 = (3025\underline{/40°})(15\underline{/-10°})$

   $\quad Z_3 = 45.375 \times 10^3 \underline{/30°}$

It is evident, from the examples shown, that the polar form is the more convenient form when one is multiplying complex quantities. Again, as we pointed out previously, each form has its own limitations.

For division as in multiplication, it also appears that the polar form rather than the rectangular form is the more convenient form.

The division of two complex numbers in rectangular form is performed in the following manner. Let $Z_1 = R_1 + jX_1$ and $Z_2 = R_2 + jX_2$. Then

$$Z_3 = \frac{Z_1}{Z_2} = \frac{R_1 + jX_1}{R_2 + jX_2}$$

In order to separate the real and imaginary parts of the phasor Z_3, it is necessary to eliminate the j part of the denominator. The procedure is to multiply both the numerator and denominator by the conjugate of the denominator. The conjugate of a complex number is found by merely changing the sign of the j or imaginary part of the rectangular form. This multiplication by unity rationalizes the fraction. Thus,

$$Z_3 = \frac{(R_1 + jX_1)(R_2 - jX_2)}{(R_2 + jX_2)(R_2 - jX_2)} = \frac{R_1 R_2 + X_1 X_2 + j(R_2 X_1 - R_1 X_2)}{R_2^2 + X_2^2}$$

and

$$Z_3 = \underbrace{\left(\frac{R_1 R_2 + X_1 X_2}{R_2^2 + X_2^2}\right)}_{R_3} + j \underbrace{\left(\frac{R_2 X_1 - R_1 X_2}{R_2^2 + X_2^2}\right)}_{X_3}$$

It should be noted that the denominator $R_2^2 + X_2^2$ of each term is a real number and, correspondingly, R_3 is real and X_3 is the imaginary part of Z_3.

Some illustrative problems will demonstrate the theory.

---

sample problem

Evaluate Z_3.

$$Z_3 = \frac{4 + j3}{1 + j2}$$

---

solution

$$Z_3 = \frac{(4 + j3)(1 - j2)}{(1 + j2)(1 - j2)}$$

$$Z_3 = \frac{4 - j8 + j3 - j^2 6}{1^2 + 2^2}$$

$$Z_3 = \frac{10 - j5}{5} = 2 - j1$$

---

sample problem

Evaluate

$$Z_3 = \frac{-2 - j4}{3 - j1}$$

solution

$$Z_3 = \frac{(-2-j4)(3+j1)}{(3-j1)(3+j1)}$$

$$Z_3 = \frac{-6-j12-j2-j^24}{3^2+1^2}$$

$$Z_3 = \frac{-2-j14}{10} = -0.2-j1.4$$

In polar form, division is accomplished by merely dividing the magnitudes of the given phasors and subtracting the angle of the denominator from the numerator. Thus,

$$Z_3 = \frac{|Z_1|\underline{/\theta_1}}{|Z_2|\underline{/\theta_2}} = \left|\frac{Z_1}{Z_2}\right|\underline{/\theta_1-\theta_2}$$

Some sample problems will demonstrate the theory.

---

sample problem

Evaluate

$$Z_3 = \frac{27\underline{/44°}}{2\underline{/7°}}$$

---

solution

$$Z_3 = \frac{27\underline{/44°}}{2\underline{/7°}} = 13.5\underline{/37°}$$

---

sample problem

Evaluate

$$Z_3 = \frac{4\underline{/130°}}{12\underline{/-15°}}$$

---

solution

$$Z_3 = \frac{4\underline{/130°}}{12\underline{/-15°}} = 0.333\underline{/145°}$$

Some concluding examples will demonstrate the interrelationship of the algebraic techniques as applied to complex numbers.

examples

Evaluate the following problems, leaving the solution in either the rectangular form or the polar form.

$$\text{(a) } Z_3 = \frac{(-2+j3)+(4+j1)}{(3+j3)-(1-j2)}$$

$$\text{(b) } Z_3 = \frac{10+j5}{20+j16} + 1\underline{/60°}.$$

$$\text{(c) } Z_3 = \frac{(5\underline{/40°})^2(1+j2)}{10\underline{/-20°}}$$

$$\text{(d) } Z_3 = \frac{(12-j5)(3+j2)}{40\underline{/-64°}+18\underline{/230°}}$$

solution

$$\text{(a) } Z_3 = \frac{(-2+j3)+(4+j1)}{(3+j3)-(1-j2)} = \frac{2+j4}{2+j5}$$

$$Z_3 = \frac{(2+j4)(2-j5)}{(2+j5)(2-j5)} = \frac{4+j8-j10-j^2 20}{2^2+5^2}$$

$$Z_3 = \frac{24-j2}{29} = 0.826-j0.012$$

$$\text{(b) } Z_3 = \frac{10+j5}{20+j16} + 1\underline{/60°}$$

$$Z_3 = \frac{(10+j5)(5-j4)}{4(5+j4)(5-j4)} + 1\underline{/60°}$$

$$Z_3 = \frac{50+j25-j40-j^2 20}{4(5^2+4^2)} + 1\underline{/60°}$$

$$Z_3 = \frac{70-j15}{164} + (\cos 60°+j \sin 60°)$$

$$Z_3 = 0.426-j0.0915+0.5+j0.866$$

$$Z_3 = 0.926+j0.7745$$

$$\text{(c) } Z_3 = \frac{(5\underline{/40°})^2(1+j2)}{10\underline{/-20°}}$$

$$Z_3 = \frac{(25\underline{/80°})(2.24\underline{/63.5°})}{10\underline{/-20°}}$$

$$Z_3 = 5.6\underline{/163.5°}$$

(d) $Z_3 = \dfrac{(12-j5)(3+j2)}{40\underline{/-64°}+18\underline{/230°}}$

$Z_3 = \dfrac{44+j9}{17.5-j36+(-11.55-j13.75)}$

$Z_3 = \dfrac{45.8\underline{/11.3°}}{5.95-j49.75}$

$Z_3 = \dfrac{45.8\underline{/11.3°}}{50\underline{/-83.18°}}$

$Z_3 = 0.916\underline{/94.48°}$

Two phasors are said to be equal if their real parts are equal and their j parts are equal. Thus, let $Z_1 = R_1+jX_1$ and $Z_2 = R_2+jX_2$. Then

$$Z_1 = Z_2 \quad \text{only if} \quad \left\{ \begin{array}{l} R_1 = R_2 \\ X_1 = X_2 \end{array} \right.$$

When a phasor is equal to zero, that is, $Z_1 = R_1+jX_1 = 0$, then R_1 must equal zero and X_1 must equal zero independently. Thus,

$$Z_1 = 0, \quad \text{then} \quad \left. \begin{array}{l} R_1 = 0 \\ X_1 = 0 \end{array} \right\} \quad \text{independently.}$$

phasors

Kirchhoff's voltage law states that at any instant of time, the algebraic sum of the various voltage drops in a circuit must equal the applied emf's. One method of summing two sinusoids of the same frequency but different phase angle is to place both sinusoidal waveforms on the same axis and then add the two waveforms point by point to yield a resultant waveform. This addition is unfortunately an extremely tedious and laborious method. The resultant sum of sine waveforms of the same frequency but different phase angle is also a sine wave of the same frequency.

Another method that can be used in summing sinusoids of the same frequency and different phase angle is called *vector addition*. The following definitions will define the nomenclature.

1. A *scalar quantity* is a quantity that has magnitude only. For example, 6 feet, 30°, and 70 miles are all scalar quantities.

2. A *vector* is a quantity having magnitude and direction. An example of a vector is: The wind velocity is 30 miles per hour in a northwesterly direction.

3. A *phasor* is a rotating vector with a constant magnitude and constant angular velocity. An example of a phasor quantity is a voltage sinusoid or a current sinusoidal waveform.

The algebraic rules applicable to vectors can also be utilized for phasor operation. Thus, note the method of representing a sinusoid as a radius vector or phasor as shown in Fig. 12–6. It is evident, therefore, that sinusoidal

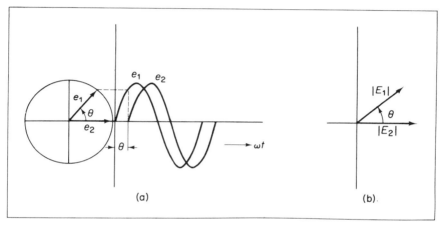

figure 12–6
(a) rotating vectors; (b) phasor diagram

waveforms of the same frequency can be represented by *fixed* phasors with the angle between the radius vectors corresponding to the phase difference between the sinusoidal waveforms.

Since the effective value of a sine wave is utilized to a much greater extent than peak, the phasor will be redefined for simplicity of circuit analysis as having a magnitude equal to the effective value. The phase angle will still have its original definition.

The method of converting from the time domain into the phasor domain is performed in the following manner. Let $e_1 = 500 \sin \omega t$ represent a voltage in the time domain. The effective value of the voltage is equal to $500/\sqrt{2}$; the phase angle is zero. Thus, in tabular form, the conversion is

Time	*Phasor*
$e = 500 \sin \omega t \quad =$	$\dfrac{500}{\sqrt{2}} \underline{/0°}$

Having converted from time to the phasor domain, we find that the techniques of vector or phasor algebra are now applicable. After the problem has been solved in the phasor domain, the solution is converted back into the time domain. Some illustrative problems will demonstrate the theory.

---

sample problem

Convert the given voltages to the phasor domain.

	Time	*Phasor*
(a) $e =$	$\sqrt{2}\ 30 \sin \omega t$	$30\underline{/0°}$
(b) $e =$	$\sqrt{2}\ 65 \sin (\omega t + 30°)$	$65\underline{/30°}$
(c) $e =$	$90 \sin (\omega t - 65°)$	$\dfrac{90}{\sqrt{2}}\underline{/-65°}$

---

sample problem

Determine the resultant voltage e.

$$e = 40 \sin (200t + 30°) + 20 \sin (200t + 60°)$$

---

solution

step 1: Let $e_1 = 40 \sin (200t + 30°)$. Convert this time function to a phasor quantity.

$$E_1 = \frac{40}{\sqrt{2}}\underline{/30°}$$

step 2: Let $e_2 = 20 \sin (200t + 60°)$. Convert this time function to a phasor quantity.

$$E_2 = \frac{20}{\sqrt{2}}\underline{/60°}$$

step 3: Convert E_1 and E_2 to rectangular form.

$$E_1 = \frac{40}{\sqrt{2}}(0.866 + j0.5)$$

$$E_2 = \frac{20}{\sqrt{2}}(0.5 + j0.866)$$

step 4:   Find resultant $E = E_1 + E_2$. Note that $20/\sqrt{2}$ is common to both terms. Thus,

$$E = \frac{20}{\sqrt{2}}[2(0.866+j0.5)+(0.5+j0.866)]$$

$$E = \frac{20}{\sqrt{2}}(2.232+j1.866)$$

step 5:   Convert E to polar form.

$$E = \frac{20}{\sqrt{2}}(2.92\underline{/39.8°})$$

$$E = \frac{58.4}{\sqrt{2}}\underline{/39.8°}$$

step 6:   Convert back into the time domain.

$$e = 58.4 \sin(200t+39.8°)$$

problems

1. Convert the following from rectangular to polar form.
(a) $3+j4$ (b) $4+j4$
(c) $14+j3.5$ (d) $7.2-j9$
(e) $-12+j5$ (f) $0.007+j0.002$
(g) $-5.3-j9$ (h) $72-j120$
(i) $-11+j3$ (j) $-18-j42$

2. Convert the following from polar to rectangular form.
(a) $8\underline{/60°}$ (b) $14.14\underline{/45°}$
(c) $135\underline{/30°}$ (d) $160\underline{/120°}$
(e) $120\underline{/-53.1°}$ (f) $20\underline{/240°}$
(g) $15\underline{/-30°}$ (h) $10\underline{/145°}$
(i) $150\underline{/320°}$ (j) $60\underline{/-230°}$

3. Perform the following operations and convert the answers to polar form.
(a) $(4.4+j7.3)+(7.6+j4.7)$
(b) $(3.65+j8.7)-(4.35+j2.7)$
(c) $6\underline{/20°}+10\underline{/60°}$
(d) $(-36+j45)-(-50+j69)$
(e) $40\underline{/-30°}-36\underline{/120°}$

4. Perform the following operations and convert the answers to polar form.
   (a) $(1+j2)(3+j5)$
   (b) $(6.5+j3.5)(2-j2)$
   (c) $(36\underline{/120°})(20\underline{/240°})+(40\underline{/-30°})(80\underline{/-60°})$
   (d) $(43\underline{/35°})(20-j35)(-40-j10)$

5. Perform the following operations and convert the answers to rectangular form.

   (a) $\dfrac{(3+j4)+(7+j6)}{(5+j3)-(3+j3)}$

   (b) $\dfrac{(1+j2)(6\underline{/30°})}{12\underline{/0°}+10+j22)}$

   (c) $\dfrac{(6\underline{/30°})(120\underline{/-60°})(5+j12)}{39\underline{/30°}}$

   (d) $\dfrac{(15\underline{/28°})(4\times10^{-3}\underline{/62°})(2+j3)}{10.8\underline{/-10°}}$

   (e) $\dfrac{(13+j39)(0.05+j0.08)(20+j40)}{(20\underline{/30°})-(40\underline{/60°})+30\underline{/45°}}$

6. Evaluate the following function
   $$Z = 4\underline{/60°}-2\underline{/30°}$$
   (a) By graphical techniques.
   (b) By vector algebra.

7. Express the following waves as phasors.
   (a) $\sqrt{2}(100)\sin(300t+30°)$
   (b) $\dfrac{\sqrt{2}}{4}\sin(400t-40°)$
   (c) $100\cos250t$
   (d) $141.4\sin(754t+0°)$
   (e) $35.35\sin(300t-20°)$

8. Express the following phasors as time functions if the frequency is 200 Hz.
   (a) $40\underline{/10°}$
   (b) $18\times10^{-3}\underline{/120°}$
   (c) $1000\underline{/-30°}$
   (d) $7.07\underline{/-90°}$
   (e) $\dfrac{400}{\sqrt{2}}\underline{/-180°}$

9. Evaluate $e = e_1 + e_2$ as a time function when
$e_1 = 60 \sin (500t - 30°)$
$e_2 = 30 \sin (500t - 10°)$

10. Evaluate $i = i_1 + i_2$ as a time function when
$i_1 = 60 \times 10^{-3} \cos 300t$
$i_2 = 20 \times 10^{-3} \sin (300t + 45°)$

11. Evaluate $e = e_1 + e_2 + e_3$ as a time function when
$e_1 = 120 \sin 300t$
$e_2 = 60 \sin (300t + 30°)$
$e_3 = 30 \sin (300t + 60°)$

12. Evaluate $i = i_1 + i_2 + i_3$ as a time function when
$i_1 = 100 \sin 200t$
$i_2 = 50 \sin (200t + 40°)$
$i_3 = 25 \sin (200t - 35°)$

ac circuits

introduction

This section will consider the analysis of sinusoidal waveforms applied to circuits that have R, X_L, and X_c as single elements and in combinations. The primary considerations in circuit problems with an applied sinusoidal emf are the magnitude and phase angle of the current that will flow.

pure R circuit

In the previous discussion, Chapter 2, it was found that for a purely resistive circuit, the voltage across R and the current flowing through R were in phase with each other. Let a sinusoidal emf be applied to the pure R series circuit shown in Fig. 13-1.

The applied emf $e = E_m \sin \omega t$ can be represented by its phasor value as

$$E = \frac{E_m}{\sqrt{2}} \underline{/0°}$$

Applying Ohm's law and some phasor algebra, we obtain

$$I = \frac{E}{R_T} = \frac{E_m}{\sqrt{2}(R_1 + R_2)} \underline{/0°}$$

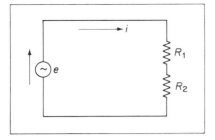

figure 13–1
ac applied to a pure R series circuit

Since the phase angle associated with resistance is zero (current and voltage are in phase with each other), then

$$I = \frac{I_m}{\sqrt{2}} \underline{/0°}$$

where

$$I_m = \frac{E_m}{R_1 + R_2}$$

Converting from a phasor quantity back to time yields

$$i = \frac{E_m}{R_1 + R_2} \sin \omega t$$

---

sample problem

The following circuit is given. The applied voltage is

$$e = 100 \sin \omega t$$

Find the value of i as a time function.

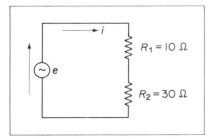

solution

step 1:   Convert e to a phasor quantity.

$$E = \frac{100}{\sqrt{2}} \underline{/0^\circ}$$

step 2:   Calculate R_T.

$$R_T = R_1 + R_2 = 40 \ \Omega$$

step 3:   Calculate I.

$$I = \frac{E}{R_T} = \frac{100\underline{/0^\circ}}{\sqrt{2}(40)}$$

$$I = \frac{2.5}{\sqrt{2}} \underline{/0^\circ}$$

step 4:   Convert I to i.

$$i = 2.5 \sin \omega t \ \text{A}$$

---

sample problem

Given $e = 60 \sin (\omega t + 30^\circ)$, find i.

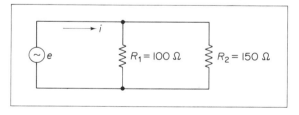

---

solution

step 1:   Convert e to E.

$$E = \frac{60}{\sqrt{2}} \underline{/30^\circ}$$

step 2:   Calculate R_T.

$$R_T = \frac{R_1 R_2}{R_1 + R_2} = \frac{(100)(150)}{100 + 150} \ \Omega$$

$$R_T = 60 \ \Omega$$

step 3:   Calculate I.

$$I = \frac{E}{R_T}$$

$$I = \frac{60\underline{/30°}}{\sqrt{2}\ 60}$$

$$I = \frac{1}{\sqrt{2}}\underline{/30°}$$

step 4:   Convert I to i.

$$i = 1\sin(\omega t + 30°)\ \text{A}$$

pure L circuit

In an L circuit, the applied voltage must be equal and opposite to the emf induced in the inductor before current will flow. It was shown that current lags the self-induced voltage by 90°. In the circuit shown in Fig. 13–2, let the self-induced voltage be equal to $e = E_m\sin\omega t$. Converting this time function to a phasor yields

$$E = \frac{E_m}{\sqrt{2}}\underline{/0°}$$

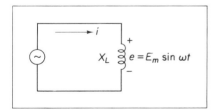

figure 13–2
ac applied to a pure L circuit

Since current lags the voltage through a coil by 90°, the instantaneous current must be equal to

$$i = I_m\sin(\omega t - 90°)$$

or in phasor form

$$I = \frac{I_m}{\sqrt{2}}\underline{/-90°}$$

Using Ohm's law relationship for the circuit yields

$$I = \frac{E_m}{\sqrt{2}\,X_L}\underline{/0 \pm \theta}$$

It is evident that $-\theta$ must equal $-90°$ and $I_m = E_m/X_L$ for the two equations to be exactly identical. Another method of writing the same equation takes into consideration the j operator. Thus,

$$I = \frac{E_m}{\sqrt{2}\,jX_L}\underline{/0°}$$

Consequently, across a pure L circuit, the current will always be in phasor form

$$I = \frac{E}{jX_L}\underline{/0°}$$

and, in time, the current will be

$$i = \frac{E_m}{X_L}\sin(\omega t - 90°)$$

---

sample problem

Determine the instantaneous current through a 0.5-H inductor when the emf across the coil is equal to $40\sin(400t + 10°)$.

---

solution

step 1:   Convert e to E.

$$E = \frac{40}{\sqrt{2}}\underline{/10°}$$

step 2:   Calculate X_L.

$$X_L = \omega L = 400(0.5) = 200\ \Omega$$

step 3:   Calculate I.

$$I = \frac{E}{jX_L} = \frac{40}{\sqrt{2}(200)}\underline{/-90°}$$

step 4:   Convert I to i.

$$i = 0.2 \sin (400t - 80°) \text{ A}$$

pure C circuit

For the pure capacitor, the current leads the voltage across the capacitor by 90°. Thus, in the circuit shown in Fig. 13–3, assume a sinusoidal voltage across C; converting to phasor form then yields

$$E \underline{/0} = E_m \sin \omega t$$

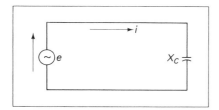

figure 13–3
ac applied to a pure C circuit

Since the current must lead the voltage across the capacitor by 90°, the equation for current must be

$$I = \frac{E}{X_c} \underline{/90°}$$

or

$$I = \frac{E}{-jX_c}$$

Note that the $+j$ is associated with an inductor and $-j$ is connected with the capacitor.

---

sample problem

Determine the instantaneous current flowing through a 0.4-μf capacitor when the voltage across it is equal to $100 \sin (500t + 10°)$.

solution

step 1: Convert e to E.

$$E = \frac{100}{\sqrt{2}} \underline{/10°}$$

step 2: Calculate X_c.

$$X_c = \frac{1}{\omega C} = \frac{1}{500(0.4) \times 10^{-6}}$$

$$X_c = 5000 \ \Omega$$

step 3: Calculate I.

$$I = \frac{E}{-jX_c} = \frac{100\underline{/10°}}{\sqrt{2}(5000)\underline{/-90°}}$$

$$I = \frac{0.02}{\sqrt{2}} \underline{/100°}$$

step 4: Convert I to i.

$$i = 0.02 \sin (500t + 100°) \text{ A}$$

series RL circuit

A circuit that contains resistance and inductance in series is shown in Fig. 13–4. The application of a voltage will cause the current to flow.

Since the current in a series circuit is the only parameter common to all elements, the reference phasor will be the current. In dc circuit theory, Kirchhoff's voltage law requires that the *algebraic* sum of all the voltages in

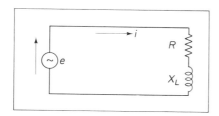

figure 13–4
series *RL* circuit

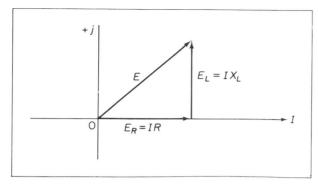

figure 13–5
phasor diagram of an *RL* circuit

a closed loop be equal to zero. In an ac circuit, the law must be revised, as the *phasor* sum of all the voltages in a closed loop must be equal to zero.

The voltage drop across R is in phase with the current, whereas the voltage drop across X_L leads the current by 90°. As shown in Fig. 13–5, the phasor voltage across R lies on the abscissa or current reference axis, and the phasor voltage across X_L lies on the $+j$ coordinate axis. Thus,

$$E = \underbrace{IR}_{E_R} + \underbrace{jIX_L}_{E_L}$$

The total opposition that a circuit presents to the flow of ac current is called the *impedance* of the circuit. The letter symbol for impedance is the capital letter Z. From the phasor diagram, an impedance triangle can be constructed as shown in Fig. 13–6.

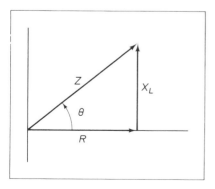

figure 13–6
impedance triangle

The magnitude of Z in this situation is equal to

$$|Z| = \sqrt{R^2 + X_L^2}$$

and the phase angle is

$$\theta = \tan^{-1} \frac{X_L}{R}$$

Since current is the dependent variable, there must be some procedure for determining the impedance of a network before evaluating the current. Consequently, we can solve for the impedance of a circuit by dividing the voltage equation by the current I. Thus,

$$\frac{E}{I} = Z = R + jX_L$$

It should be noted that although it appears that impedance has both magnitude and direction, the impedance of a circuit is a scalar quantity and a nonvariant function with respect to time. The relationship among resistance, reactance, and impedance is numerically identical to the relationship between the applied voltage and the voltage drops in the circuit. Therefore, the impedance of a series circuit can be calculated by the vector sum of the resistance and reactance. An illustrative problem will demonstrate the theory.

---

sample problem

A 0.5-H inductor and a resistor of 300 ohms are connected in series across a line having a voltage of 100 sin (600t). What are the magnitude and phase of the current with respect to the applied emf?

---

solution

step 1: Calculate X_L.

$$X_L = \omega L = 600 \times 0.5 = 300 \ \Omega$$

step 2: Construct the impedance triangle. Calculate Z.

$$|Z| = \sqrt{R^2 + X_L^2} = \sqrt{300^2 + 300^2}$$
$$|Z| = 424.2 \ \Omega$$

$$\theta = \tan^{-1} \frac{X_L}{R} = \tan^{-1} \frac{300}{300}$$

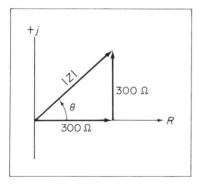

$$\theta = 45°$$
$$Z = 424.2 \underline{/45°} \ \Omega$$

step 3:   Calculate *I*.

$$I = \frac{E}{Z} = \frac{70.7 \underline{/0°}}{424.2 \underline{/45°}}$$

$$I = 0.167 \underline{/-45°} \ A$$

---

sample problem

In the given circuit, the inductor and resistor comprise an impedance of $100 \underline{/60°}$ ohms. An external resistor is connected to the impedance. Determine:

(a) The coil inductance when $\omega = 1000$ rad/sec.
(b) The current in the circuit.
(c) The voltage drop across each element.
(d) The phasor diagram.

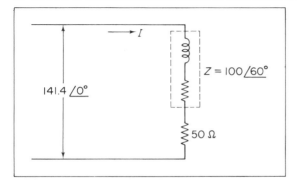

solution

step 1:   Convert Z to rectangular form.

$$Z = 100(\cos 60° + j \sin 60°)$$
$$R = 50 \ \Omega$$
$$X_L = 86.6 \ \Omega$$

step 2:   Calculate L.

$$L = \frac{X_L}{\omega} = \frac{86.6}{1000}$$
$$L = 86.6 \text{ mH.}$$

step 3:   Calculate the total circuit impedance.

$$Z_T = 50 + (50 + j86.6) = 100 + j86.6 = 130 \underline{/40.9°}$$

step 4:   Solve for the current I.

$$I = \frac{E}{Z_T} = \frac{141.4 \underline{/0°}}{130 \underline{/40.9°}}$$
$$I = 1.085 \underline{/-40.9°}$$

step 5:   Calculate the voltage drops.

$$E_{50} = I(50) = 1.085 \times 50 = 54.25 \text{ v}$$
$$E_{50} = 54.25 \text{ v}$$
$$E_{86.6} = 1.085 \times 86.6 = 94 \text{ v}$$

step 6:   The phasor diagram is

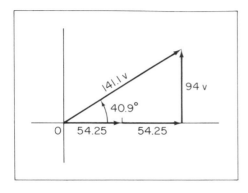

A pure inductor consumes no power; therefore, all the power dissipated in an RL series circuit (Fig. 13–4) must be accounted for by the resistor. Thus, the average power dissipated in the circuit is

$$P = I^2R = I(IR)$$

Note that $IR = E_R = E \cos \theta$. Then

$$P = EI \cos \theta \text{ w}$$

This power dissipated in the resistor is called the *true power*. The power stored in the magnetic field of the inductor is called the *reactive power*. The product of the voltage across the inductor and the current flowing through the inductor is equal to the reactive power. The letter symbol for the power across an inductor is P_L. To prevent any error between dissipated and the stored power, the unit of reactive power is the voltampere, abbreviated var. The input power required to supply the circuit is called the *apparent power*. Thus,

$$P_a = P_R + jP_L$$

or

$$P_a = \sqrt{P_R^2 + P_L^2} \text{ vars}$$

power factor

A general definition of power factor that is valid for all ac and non-sinusoidal waveforms is the ratio of the real power output to the apparent power input. Thus,

$$\text{Power factor} = \text{P.F.} = \frac{\text{true power (watts)}}{\text{apparent power (vars)}}$$

For sinusoidal waveforms only,

$$\text{P.F.} = \cos \theta = \frac{EI \cos \theta \text{ (watts)}}{EI \text{ (vars)}}$$

Note that the power factor can be determined by evaluating the cosine of the phase angle existent between the input voltage and the current. The angle θ is called the power factor angle of the circuit. The impedance triangle can also be used to establish the value of θ. Thus,

$$\cos \theta = \frac{R}{|Z|} = \frac{R}{\sqrt{R^2 + X_L^2}}$$

It is evident that the power factor is a numerical quantity varying between zero and unity. Note that a unity power factor denotes a purely resistive network, whereas a zero power factor denotes a purely reactive network.

An illustrative problem will demonstrate the theory.

---

sample problem

A series RL circuit has an impedance of $1000\underline{/60°}$ ohms to a 300-volt, 500-Hz source of emf. Find: (a) R and L, (b) I, (c) P, (d) P.F.

---

solution

step 1:   Calculate R and L.

$$Z = 1000(\cos 60° + j \sin 60°)$$
$$R = 500 \ \Omega$$
$$X_L = 866 \ \Omega$$
$$L = \frac{X_L}{\omega} = \frac{866}{2\pi(500)} = 0.276 \ \text{H}$$

step 2:   Calculate I.

$$I = \frac{E}{Z} = \frac{212\underline{/0°}}{1000\underline{/60°}}$$
$$I = 0.212\underline{/-60°} \ \text{A}$$

step 3:   Calculate P.

$$P = I^2R = (0.212)^2 500$$
$$P = 22.5 \ \text{w}$$

A check on this solution uses the basic equation for power. Thus,

$$P = EI \cos \theta = 212 \times 0.212 \cos 60°$$
$$P = 22.5 \ \text{w}$$

step 4:   Calculate the P.F.

$$\text{P.F.} = \cos \theta = \cos 60° = 0.5 \ \text{lagging}$$

series RC circuit

A resistor is connected in series with a capacitor across an ac circuit source as shown in Fig. 13–7. Again, in a series circuit, the current is common

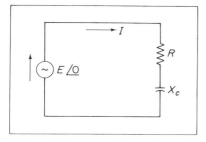

figure 13–7
series *RC* circuit

to all circuit elements. The voltage across the resistor is in phase with the current, whereas the voltage across the capacitor lags the current by 90°. The phasor voltage across *R* lies on the abscissa or current reference axis, whereas the phasor voltage across X_c lags the current by 90° and lies on the $-j$ axis, as shown in Fig. 13–8.
Thus,

$$E = \underbrace{IR}_{E_R} - \underbrace{jIX_c}_{E_c}$$

The magnitude of the impedance of an *RC* series circuit is

$$|Z| = \sqrt{R^2 + X_c^2}$$

and the phase angle is

$$\theta = \tan^{-1}\frac{-X_c}{R}$$

Note that

$$Z = |Z|\underline{/-\theta} = R - jX_c$$

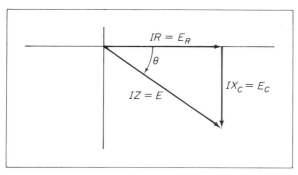

figure 13–8
phasor diagram for a series *RC* circuit

It is evident that in rectangular coordinates, inductive reactance is always a $+j$ quantity, whereas capacitive reactance is always a $-j$ quantity. The impedance of a series RL circuit usually is in the first quadrant and has a phase angle variant between 0 and 90°. The impedance of a series RC circuit is usually in the fourth quadrant and has a phase angle variant between 0 and $-90°$.

An illustrative problem will demonstrate the theory.

---

sample problem

Given the following circuit, determine:

        (a) The total impedance in polar form.
        (b) The current I.
        (c) The voltage drop across each element.

Also, construct the phasor diagram.

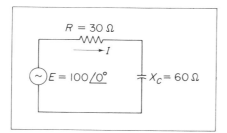

solution

step 1:   Calculate Z_T.

$$Z_T = 30 - j60 = 67.2\underline{/-63.5°}$$

step 2:   Calculate I.

$$I = \frac{E}{Z_T} = \frac{100\underline{/0°}}{67.2\underline{/-63.5°}}$$

$$I = 1.485\underline{/63.5°}$$

step 3:   Calculate the voltage drops.

$$E_R = IR = 1.485 \times 30$$
$$E_R = 44.55 \text{ v}$$
$$E_c = 1.485 \times 60 = 89.1 \text{ v}$$

step 4:   Construct the phasor diagram. Note that $E = E_R - jE_c$ or $E = \sqrt{E_R^2 + E_c^2}$.

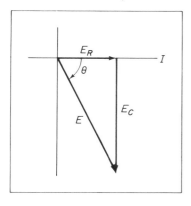

power considerations

A pure capacitor consumes no power; therefore, all the power dissipated in a series RC circuit (Fig. 13–7) must be accounted for by the resistor. Thus, the average power dissipated in the circuit is

$$P = I^2 R = EI \cos \theta \text{ w}$$

The power dissipated in the resistor is called the true power. The power stored in the electrostatic field of the capacitor is called the reactive power. The letter symbol for the power across the capacitor is P_c.

The apparent power is

$$P_a = P_R - jP_c$$

The magnitude of the apparent power is

$$P_a = \sqrt{P^2 + P_c^2} \text{ vars}$$

The power factor is defined as the ratio of the true power to the apparent power. For sinusoidal waveforms only, the value is

$$\cos \theta = \frac{EI \cos \theta}{EI}$$

or

$$\cos \theta = \frac{R}{|Z|} = \frac{R}{\sqrt{R^2 + X^2}}$$

The power factor in a series RC circuit usually is considered a leading phase angle, whereas in a series RL circuit, the power factor is a lagging phase angle.

An illustrative problem will demonstrate the theory.

---

sample problem

A series RC circuit has an impedance equal to $500/-60°$ ohms to a 500-volt, 300-Hz source emf. Find: (a) R and C, (b) I, (c) P_R, (d) P.F.

---

solution

step 1:   Convert Z to rectangular form.

$$Z = 500(\cos \theta - j \sin \theta)$$
$$Z = 500(\cos 60° - j \sin 60°)$$
$$Z = 250 - j433 \ \Omega$$
$$R = 250 \ \Omega$$
$$X_c = 433 \ \Omega$$
$$C = \frac{1}{2\pi f X_c} = \frac{1}{2\pi \times 300 \times 433}$$
$$C = 0.398 \ \mu f$$

step 2:   Calculate I.

$$I = \frac{E}{Z} = \frac{500}{\sqrt{2}\ 500/-60°}$$
$$I = 0.707/+60°$$

step 3:   Calculate P_R.

$$P_R = I^2 R = (0.707)^2 250 = 125 \ w$$

or, to check the results,

$$P_R = EI \cos \theta = \frac{500}{\sqrt{2}}(0.707) \cos 60°$$
$$P_R = 125 \ w$$

step 4:   Calculate P.F.

$$P.F. = \cos \theta = \cos 60° = 0.5 \text{ leading}$$

series circuits containing more than two elements

Let n coils having both resistance and inductance be connected in series as shown in Fig. 13–9. The impedance of this circuit can be determined quite readily, since the components are directly additive in a series circuit. Thus,

$$Z_T = Z_1 + Z_2 + Z_3 + \cdots + Z_n$$

or

$$Z_T = (R_1 + jX_1) + (R_2 + jX_2) + (R_3 + jX_3) + \cdots + (R_n + jX_n)$$

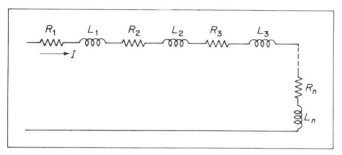

figure 13–9
series circuit containing more than two elements

Collecting the terms yields

$$Z_T = \underbrace{(R_1 + R_2 + R_3 + \cdots + R_n)}_{R_0} + j\underbrace{(X_1 + X_2 + X_3 + \cdots + X_n)}_{X_0}$$

It is evident that the total impedance of a series circuit is the sum of the resistance and the sum of the reactances, yielding

$$Z_T = R_0 + jX_0 = R_0 + j\omega L_0$$

This relationship is shown in the diagram of Fig. 13–10.

The magnitude of the total impedance is given by

$$|Z_T| = \sqrt{R_0^2 + X_0^2}$$

and the phase angle is

$$\theta_0 = \tan^{-1} \frac{X_0}{R_0}$$

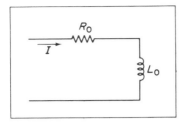

figure 13–10
equivalent series RL circuit

The power factor of the entire circuit is

$$\text{P.F.} = \cos \theta_0 = \frac{R_0}{|Z_T|}$$

where the angle θ_0 is a lagging power factor angle. The total power consumed by the circuit is

$$P = I^2 R_0 = EI \cos \theta_0 \text{ w}$$

It is obvious that the foregoing theory can also be applied to resistors connected in series with capacitors as shown in Fig. 13–11. Using the previous theory of series impedances yields

$$Z_T = Z_1 + Z_2 + Z_3 + \cdots + Z_n$$

or

$$Z_T = (R_1 - jX_1) + (R_2 - jX_2) + (R_3 - jX_3) + \cdots + (R_n - jX_n)$$

Collecting the terms properly yields

$$Z_T = \underbrace{(R_1 + R_2 + R_3 + \cdots + R_n)}_{R_0} - j\underbrace{(X_1 + X_2 + X_3 + \cdots + X_n)}_{X_0}$$

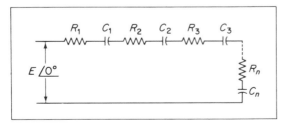

figure 13–11
series circuit containing RC elements

The equivalent relationship is shown in the circuit of Fig. 13–12. The magnitude of the total impedance is

$$|Z_T| = \sqrt{R_0^2 + X_0^2}$$

and the phase angle is

$$\theta_0 = \tan^{-1} \frac{-X_0}{R_0}$$

The power factor of the entire circuit is

$$\text{P.F.} = \cos \theta_0 = \frac{R_0}{|Z_T|}$$

where the angle θ_0 is a leading power factor angle. The total power consumed by the circuit is

$$P = I^2 R_0 = EI \cos \theta_0 \text{ w}$$

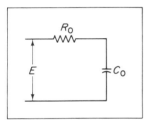

figure 13–12
equivalent *RC* series circuit

series RLC circuit

Figure 13–13 shows a circuit containing three series elements, namely, resistance, inductance, and capacitance in series. The current is the same throughout the circuit elements. The voltage across the resistance is in phase with the current. The voltage across the inductor leads the current by 90°, and the voltage across the capacitor lags the current by 90°. The voltage drops across the inductance and capacitance, respectively, are 180° out of phase. They oppose each other, and the resultant voltage is the arithmetic difference of the two voltages.

Kirchhoff's law equation for the voltages in a series circuit can be expressed by

$$E = IR + jX_L I - jIX_c$$

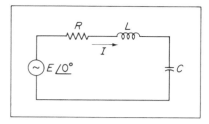

figure 13–13
series *RLC* circuit

Note that

$$\frac{E}{I} = Z = R + j(X_L - X_c)$$

The applied emf will either lead or lag the current, depending upon whether X_L is greater than or smaller than X_c. When X_L is greater than X_c, the circuit is inductive in nature, and the applied emf leads the current. When X_L is smaller than X_c, the circuit is capacitive in nature and the applied voltage E lags the current. The power factor angle is positive in an inductive circuit and negative in a capacitive circuit.

Consider the phasor diagram shown in Fig. 13–14 with X_L greater than X_c. The magnitude of impedance Z is

$$|Z| = \sqrt{R^2 + (X_L - X_c)^2}$$

and the phase angle is

$$\theta = \tan^{-1}\left(\frac{X_L - X_c}{R}\right)$$

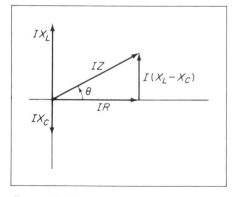

figure 13–14
phasor diagram with X_L greater than X_c

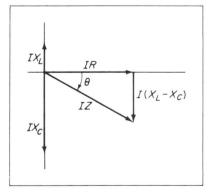

figure 13–15
phasor diagram with X_L smaller than X_c

In the phasor diagram shown in Fig. 13–15, X_L is smaller than X_c. The magnitude of impedance is

$$|Z| = \sqrt{R^2 + (X_L - X_c)^2}$$

and the phase angle is

$$\theta = \tan^{-1}\left(\frac{X_L - X_c}{R}\right)$$

When X_L is greater than X_c, the circuit may be considered as equivalent to a simple RL series circuit, whereas when X_L is smaller than X_c, the circuit appears as a series RC circuit.

The power factor of a series RLC circuit is

$$\text{P.F.} = \cos\theta = \frac{R}{\sqrt{R^2 + (X_L - X_c)^2}}$$

The power consumed in the circuit is

$$P = I^2 R = EI\cos\theta \text{ w}$$

An illustrative problem will demonstrate the theory.

---

sample problem

A capacitor having a reactance of 120 ohms is connected in series with a coil having a resistance of 60 ohms and a reactance of 180 ohms. The voltage input is a 200-volt, 1000-Hz source. Find: (a) Z, (b) I, (c) P.F., (d) P, (e) the phasor diagram.

---

solution

step 1: Calculate Z.

$$Z = R + j(X_L - X_c) = 60 + j(180 - 120)$$
$$Z = 60 + j60 = 84.84\underline{/45°}\ \Omega$$

step 2: Calculate I.

$$I = \frac{E}{Z} = \frac{200\underline{/0°}}{84.84\underline{/45°}}$$
$$I = 2.36\underline{/-45°}\ \text{A}$$

step 3: Calculate P.F.

$$\cos \theta = \cos 45° = 0.707 \text{ lagging P.F.}$$

step 4: Calculate P.

$$P = I^2R = (2.36)^2 60 = 333 \text{ w}$$

step 5: Construct the phasor diagram.

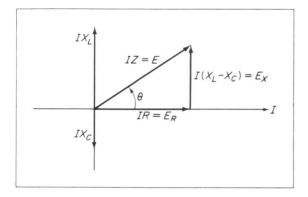

problems

1. The reactance of a coil is 200 ohms and its inductance is 500 mH. Determine its impedance, its current, the power factor, and the total power consumed by the circuit when the coil is connected across a 100-volt, 1000-Hz source.

2. A coil having resistance and reactance is connected across a 200-volt, 500-Hz source, and the power consumed is 20 watts. If the magnitude of the impedance of the coil is 1000 ohms, determine the current, the power factor, and the values of R and L of the coil.

3. A coil having a resistance of 100 ohms also has a power factor of 0.866 when connected across a 100-volt, 1000-Hz source. Determine the line current, the inductance of the coil, and the power consumed in the circuit.

4. A coil is connected across a 100-volt, 1000-Hz source, and a current of 25 milliamperes flows. If the power factor is 0.765 lagging, determine the reactance and inductance of the coil and the power consumed.

5. A coil has a resistance of 750 ohms and an inductance of 159 mH. A source emf of 300 volts, 1000 Hz is connected across the coil. Determine the current, the power factor, the power consumed, the reactive power, and the apparent power.

6. In the circuit shown, the voltage across the inductor is 80 volts at 4 kHz. Find: (a) I, (b) E, (c) P.F., (d) P.

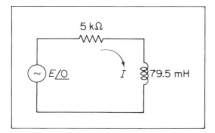

7. The impedance of a coil is $50 + j120$ ohms at a frequency of 40 kHz. The current through the coil is 20 ma. Find: (a) the applied voltage, (b) P.F., (c) P.

8. A coil has a resistance of 150 ohms and an inductance of 0.5 H. When the applied voltage is 180 volts, the current is 600 ma. Find: (a) frequency, (b) P.F., (c) P.

9. A resistor and a capacitor are connected in series across a 200-volt, 1000-Hz line. The impedance of the circuit is 500 ohms, and the power consumed is 50 watts. Find: (a) R, (b) C, (c) I, (d) P.F.

10. A coil is connected to a 24-volt source dc and draws .025 ampere. When the coil is connected to a 200-volt, 1000-Hz source, the current is 0.1 ampere. Find: (a) P.F., (b) R and L.

11. When a coil is connected to a 200-volt, 1000-Hz source, the current is .05 ampere. If the frequency is increased to 3000 Hz and the voltage remains constant, the current drops to 0.025 ampere. Find the values of R and L for the coil.

12. A 0.25-μf capacitor in series with a 1000 ohms resistance is connected across a 200-volt, 1000-Hz source. Find: (a) Z, (b) I, (c) P.F., (d) P.

13. A capacitor when connected in series with a 200-ohm resistor has an impedance of 300 ohms. A 200-volt, 500-Hz source is supplying the energy. Find: (a) C, (b) I, (c) P.F., (d) P.

14. A 500-ohm resistor is connected in series with a capacitor across a 100-volt, 2000-Hz source. If the power factor is 0.94, determine: (a) Z, (b) C, (c) I, (d) P.

15. The impedance of a series-connected RC circuit is 500 ohms at 1000 Hz. If the power factor is 0.643 and the power consumed is 80 watts, determine: (a) I, (b) E, (c) R and C.

16. The impedance of a series RC circuit is $200 - j300$ ohms. The voltage across the circuit is 100 volts–50 kHz. Determine: (a) I, (b) P.F., (c) C, (d) P.

17. A 1000-ohm resistor is in series with a capacitor across a 500-volt, 20-kHz source. The power is 50 watts. Find: (a) I, (b) C, (c) P.F.

18. Given the circuit shown, if the angular velocity applied is $\omega = 10^6$ rad/sec find: (a) Z, (b) I, (c) P.F., (d) P, (e) phasor diagram.

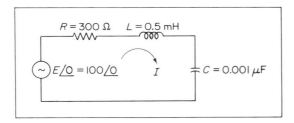

19. Given the circuit shown, find: (a) Z, (b) I, (c) P.F., (d) P, (e) phasor diagram.

$R = 100\ \Omega$ $\qquad \omega = 10^5$ rad/sec
$X_L = 300\ \Omega$ $\qquad E = 100\underline{/0°}$
$X_c = 100\ \Omega$

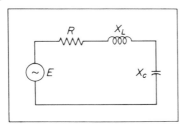

20. For the following given circuit, find: (a) Z, (b) I, (c) P.F., (d) P, (e) phasor diagram.

$R = 1000\ \Omega$ $\qquad L_2 = 300$ mH
$L_1 = \quad 500$ mH $\qquad C = 0.1\ \mu f$

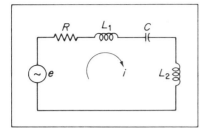

$$e = \sqrt{2}\ 200 \sin (5000t + 20°)$$

21. A series circuit of a 250-ohm resistor, a 300-μH inductor, and a capacitor are connected to a 200-volt, $10^6/\pi$-Hz source. Find: (a) the value of C required to have the current in the circuit equal to 0.5 ampere at a lagging P.F.; (b) the same current as (a) but a leading P.F. angle.

22. A series circuit consists of a 300-ohm resistor, a 200-mH inductor, and a 0.5-μf capacitor. The circuit is connected to a 300-volt source. Find, when the current in the circuit is 0.75 ampere, the frequency where the power factor angle is (a) leading, (b) lagging.

14

parallel circuits

resistors in parallel

Two impedances are said to be in parallel when the same voltage is across each of them. Note that in a series circuit, the current is the parameter that is common to all circuit elements. In parallel circuits, the voltage is the common parameter to all the circuit elements.

The analysis of the parallel combination of two resistors in parallel as shown in Fig. 14-1 is similar to the dc condition. Thus, the source current I is the phasor sum of I_1 and I_2 according to Kirchhoff's current law.

By Ohm's law, note that I_1 and I_2 have the following values.

$$I_1 = \frac{E}{R_1} \quad \text{and} \quad I_2 = \frac{E}{R_2}$$

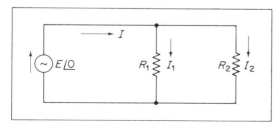

figure 14-1
two resistors in parallel with ac applied

Since the current is in phase with the voltage across a resistor, the source current is

$$I = I_1 + I_2$$

Replacing the parallel combination of two resistors by one resistor R_T results in the following value for I.

$$I = \frac{E}{R_T} = \frac{E}{R_1} + \frac{E}{R_2}$$

Note that the value of R_T is

$$\frac{1}{R_T} = \frac{1}{R_1} + \frac{1}{R_2}$$

or the solution can be simplified to the familiar product over sum rule as

$$R_T = \frac{R_1 R_2}{R_1 + R_2}$$

From previous theory, the *conductance* symbolized by G of a resistive element was defined as the reciprocal of that element. Thus, the conductance of R_1 is $G_1 = 1/R_1$, the conductance of R_2 is $G_2 = 1/R_2$, etc. The unit of conductance is the reciprocal of an ohm, or the *mho*. The conductance of a parallel circuit is equal to the sum of the conductances, or

$$G_T = G_1 + G_2 \text{ mhos}$$

The source current would then be equal to

$$I = G_T E \text{ A}$$

parallel RC circuit

A resistor R is connected in parallel with C across an ac source as shown in Fig. 14–2. In this circuit, the voltage across the parallel combination is the applied voltage. The current I is equal to the phasor sum of the branch currents I_R and I_c.

Assume $E/\underline{0°}$ to be the reference phasor. The current through R is in phase with the applied voltage, whereas the current through C leads the voltage by 90°. Thus, the source current I leads the source voltage E by a phase angle θ, as shown in Fig. 14–3.

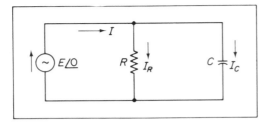

figure 14–2
parallel *RC* circuit

The value of θ can be determined by the relationship

$$\theta = \tan^{-1}\left(\frac{I_c}{I_R}\right)$$

analysis

The values of I_R and I_c are, respectively,

$$I_R = \frac{E}{R} \quad \text{and} \quad I_c = \frac{E}{-jX_c}$$

The current I is the phasor sum of the currents, or

$$I = I_R + I_c = \frac{E}{Z}$$

The total impedance of the circuit is

$$\frac{1}{Z} = \frac{1}{R} + \frac{1}{-jX_c}$$

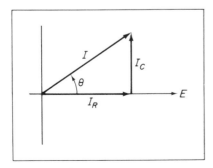

figure 14–3
phasor diagram of an *RC* parallel circuit

Using the product over sum rule results in

$$Z = \frac{(R)(-jX_c)}{R-jX_c}$$

It is evident that this value is normally rationalized by multiplying numerator and denominator by the conjugate of the denominator. The resultant value for impedance is

$$Z = \underbrace{\left(\frac{RX_c^2}{R^2+X_c^2}\right)}_{R_0} - j\underbrace{\left(\frac{R^2X_c}{R^2+X_c^2}\right)}_{X_0}$$

Note that this form of impedance is a series equivalent circuit, where R_0 is the equivalent overall series resistance and X_0 is the equivalent series reactance of the circuit. The negative sign preceding the j term denotes that the reactance of the circuit is capacitive.

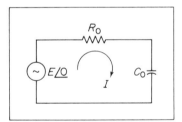

figure 14–4
equivalent series RC circuit

The phase angle for the circuit is equal to

$$\theta = \tan^{-1}\left(\frac{-R}{X_c}\right)$$

The series circuit that is equivalent to the parallel circuit is shown in Fig. 14–4.

An illustrative problem will demonstrate the theory.

sample problem

A 2000-ohm resistor is connected in parallel across a 100-pf capacitor. The source voltage is 50 volts—$10^6/\pi$ Hz. Find: (a) Z and (b) I. (c) Construct the phasor diagram.

solution

step 1:   Calculate X_c and Z.

$$X_c = \frac{1}{\omega C} = \frac{1}{2\pi \times \dfrac{10^6}{\pi} \times 100 \times 10^{-12}}$$

$$X_c = 5000 \ \Omega$$

$$Z = \frac{RX_c^2}{R^2 + X_c^2} - j\frac{R^2 X_c}{R^2 + X_c^2}$$

$$Z = \frac{2 \times 25 \times 10^9}{(4+25)10^6} - j\frac{4 \times 5 \times 10^9}{(4+25)10^6}$$

$$Z = 1720 - j690 \ \Omega$$

$$Z = 1855\underline{/-21.8^\circ} \ \Omega$$

step 2:   Calculate I.

$$I = \frac{E}{Z} = \frac{50\underline{/0^\circ}}{1855\underline{/-21.8^\circ}}$$

$$I = 27\underline{/21.8^\circ} \ \text{ma}$$

step 3:   Construct the phasor diagram.

$$I_R = \frac{E}{R} = \frac{50}{2 \times 10^3} = 25 \ \text{ma}$$

$$I_c = \frac{E}{X_c} = \frac{50}{5 \times 10^3} = 10 \ \text{ma}$$

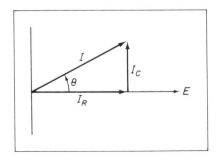

conductance, susceptance, and admittance

In dealing with impedances in parallel, it is sometimes simpler to operate with inverse impedance relationships. Consider the parallel *RC* network shown in Fig. 14-5. Using the impedance method, we find that the value of the impedance is

$$\frac{1}{Z} = \frac{1}{R} + \frac{1}{-jX_c}$$

The reciprocal of a pure resistor is defined as a *conductance*, and the letter symbol of conductance is G. The unit of conductance is the mho, as defined by the equation

$$G = \frac{1}{R}$$

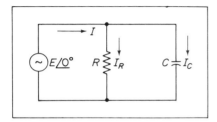

figure 14–5
parallel *RC* circuit

The reciprocal of a pure reactance is defined as a *susceptance*. The letter symbol for susceptance is the letter B. The unit of susceptance is the mho, as defined by

$$B = \frac{1}{X}$$

The reciprocal of an impedance is defined as an *admittance*. The letter symbol for admittance is the letter Y. The unit of admittance is the mho, as defined by

$$Y = \frac{1}{Z}$$

Note that the equation for the *RC* parallel circuit can be rewritten in terms of reciprocal quantities as

$$Y = G + jB_c$$

It is evident that the values of I_R and I_c are

$$I_R = GE \quad \text{and} \quad I_c = B_c E$$

Thus,

$$I = YE = (G + jB_c)E$$

A problem sometimes arises when more than two elements are connected in parallel. Consider n parallel impedances across a line with a common voltage applied as shown in Fig. 14–6.

The current I is equal to the phasor sum of currents. Thus,

$$I = I_1 + I_2 + I_3 + \cdots + I_n$$

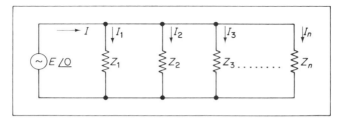

figure 14–6
n parallel impedances across a line

Since E is common to all circuit elements, the admittance of the circuit is

$$I = EY_T = E(Y_1 + Y_2 + Y_3 + \cdots + Y_n)$$

or

$$Y_T = Y_1 + Y_2 + Y_3 + \cdots + Y_n$$

where

$$Y_1 = \frac{1}{Z_1}, \qquad Y_2 = \frac{1}{Z_2}$$

$$Y_3 = \frac{1}{Z_3}, \qquad Y_n = \frac{1}{Z_n}$$

It is evident that admittances in parallel are directly additive as resistors in series. The converse is also true; that is, admittances in series are treated in a manner similar to resistors in parallel.

An admittance triangle can be developed illustrating the phase angle and magnitude of admittance in a similar manner to that of impedance.

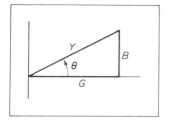

figure 14–7
admittance triangle of a parallel *RC* circuit

Thus, refer to Fig. 14–7, which is the admittance triangle of a parallel *RC* circuit.

An illustrative problem will demonstrate the theory.

---

sample problem

A 2000-ohm resistor is connected across a capacitor having a reactance of 5000 ohms. Determine, when the source voltage is 50 volts, (a) Y_T; (b) I.

---

solution

step 1:   Calculate Y_T.

$$G = \frac{1}{R} = \frac{1}{2000} = 0.5 \times 10^{-3} \text{ mho}$$

$$B = \frac{1}{X_c} = \frac{1}{5000} = 0.2 \times 10^{-3} \text{ mho}$$

$$Y_T = G + jB = (0.5 + j0.2)10^{-3} \text{ mho}$$

or

$$Y_T = 0.54\underline{/21.8°}(10^{-3}) \text{ mho}$$

step 2:   Calculate I.

$$I = EY_T = (50\underline{/0°})(0.54\underline{/21.8°})(10^{-3})$$

$$I = 27\underline{/21.8°} \text{ ma}$$

parallel RL circuit

A voltage source is connected across a resistor in parallel with an inductor as shown in Fig. 14–8. The branch current I_R is in phase with the applied voltage, whereas the current I_L lags the applied voltage by 90°. The phasor sum of the currents lags the applied voltage by some phase angle θ.

The first step in the analysis of the parallel RL circuit is to evaluate the branch currents I_R and I_L. Thus,

$$I_R = \frac{E}{R}, \qquad I_L = \frac{E}{jX_L}$$

The source current I is equal to the phasor sum of the branch currents. Thus,

$$I = I_R + I_L = \frac{E}{R} + \frac{E}{jX_L}$$

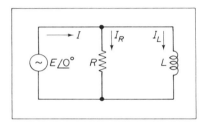

figure 14–8
parallel RL circuit

Since $I = E/Z$, then the total impedance of the circuit is

$$\frac{1}{Z} = \frac{1}{R} + \frac{1}{jX_L}$$

Simplifying the impedance equation results in

$$Z = \frac{R(jX_L)}{R + jX_L}$$

Rationalizing the expression yields

$$Z = \underbrace{\left(\frac{RX_L^2}{R^2 + X_L^2}\right)}_{R_0} + j\underbrace{\left(\frac{R^2 X_L}{R^2 + X_L^2}\right)}_{X_0}$$

The value of Z is now in the form of $Z = R_0 + jX_0$, where R_0 is the equivalent series resistance and X_0 is the equivalent series reactance. The positive sign preceding the j term denotes that the reactance is inductive. The equivalent series circuit determined from the original parallel circuit is shown in Fig. 14–9.

The phase angle of the circuit is defined as the angle whose tangent is the ratio of the reactive term to the resistive term. Thus,

$$\theta = \tan^{-1}\left(\frac{R^2 X_L}{RX_L^2}\right) = \tan^{-1}\left(\frac{R}{X_L}\right)$$

The theory of admittances can also be applied as a mathematical technique to the analysis and solution of a parallel RL circuit. Thus, the impedance of a parallel RL circuit is

$$\frac{1}{Z} = \frac{1}{R} - j\frac{1}{X_L}$$

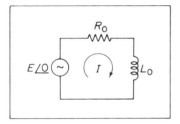

figure 14–9
series equivalent circuit

The admittance is

$$Y = G - jB_L$$

Note that the phase angle θ is

$$\theta = \tan^{-1}\left(\frac{-B_L}{G}\right)$$

Thus, the polar form of admittance is

$$Y = \sqrt{G^2 + B_L^2}\ \underline{/-\theta}$$

An illustrative problem will demonstrate the theory.

sample problem

A 500-ohm resistor is connected in parallel with a 300-ohm inductive reactance. A voltage source of 100 $\underline{/0°}$ volts is connected across the parallel combination. Determine: (a) Z, (b) Y, (c) I.

solution

step 1: Calculate Z.

$$Z = \frac{R X_L^2}{R^2 + X_L^2} + j \frac{R^2 X_L}{R^2 + X_L^2}$$

$$Z = \frac{(5)(9)10^6}{(25+9)10^4} + j \frac{(25)(3)10^6}{(25+9)10^4}$$

$$Z = 132.5 + j221 \ \Omega$$

$$Z = 257 \underline{/59°} \ \Omega$$

The impedance triangle is

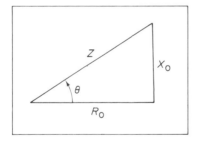

step 2: Calculate Y.

$$Y = G - jB_L$$
$$G = \tfrac{1}{500} = 2 \times 10^{-3}$$
$$B_L = \tfrac{1}{300} = 3.33 \times 10^{-3}$$
$$Y = (2 - j3.33)10^{-3}$$
$$Y = 3.88 \times 10^{-3} \underline{/-59°} \text{ mhos}$$

step 3: Calculate I.

$$I = EY = 100 \times 3.88 \times 10^{-3} \underline{/-59°}$$
$$I = 0.388 \underline{/-59°} \text{ A}$$

step 4: Construct the admittance triangle and the phasor diagram.

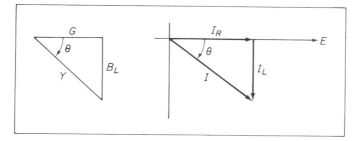

impedance and admittance

It was noted for a simple parallel circuit that the impedance could be expressed in the general form

$$Z = R_0 \pm jX_0$$

The series equivalent circuit for a given parallel circuit is shown in Fig. 14–10.

The subscripts p and o denote parallel and overall values. Thus, the R_p's and the R_o's are not equal but represent general parallel and series resistors. Note that for the two circuits to be equivalent, the same current must flow, if it is assumed that the same input voltage is applied.

The conversion formulas required to convert from a parallel circuit to an equivalent series circuit are:

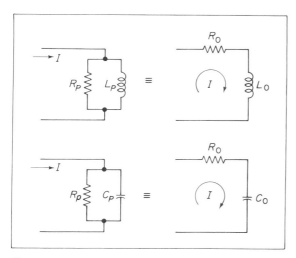

figure 14–10
equivalent series circuit

1. *RC parallel circuit:*

$$R_o = \frac{R_p X_c^2}{R_p^2 + X_c^2}$$

$$X_o = \frac{R_p^2 X_c}{R_p^2 + X_c^2}$$

2. *RL parallel circuit:*

$$R_o = \frac{R_p X_L^2}{R_p^2 + X_L^2}$$

$$X_o = \frac{R_p^2 X_L}{R_p^2 + X_L^2}$$

It is evident from the foregoing that the converse must also be valid. An equivalent parallel circuit must exist for a given series circuit. The procedure is to evaluate the total admittance of a series circuit in rectangular form and then rationalize the denominator to produce a rectangular form of admittance. Thus, the general approach for the series circuit is shown in Fig. 14–11.

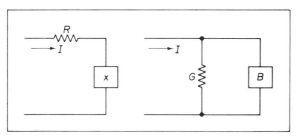

figure 14–11
parallel equivalent circuit

The impedance of a series circuit is

$$Z = R \pm jX$$

The admittance of the circuit is

$$Y = \frac{1}{Z} = \frac{1}{R \pm jX} = \frac{R \mp jX}{R^2 + X^2}$$

and

$$Y = \underbrace{\left(\frac{R}{R^2 + X^2}\right)}_{G} \mp j \underbrace{\left(\frac{X}{R^2 + X^2}\right)}_{B}$$

Consequently, to convert from parallel to series circuitry, the impedance formula is used, whereas to convert from series to parallel, the admittance technique is used.

The conversion formulas required to convert from a series circuit to an equivalent parallel circuit are

1. *Series RC circuit:*

$$G = \frac{1}{R_p} = \frac{R}{R^2 + X_c^2}$$

$$B_c = \frac{1}{X_p} = \frac{X_c}{R^2 + X_c^2}$$

2. *Series RL circuit:*

$$G = \frac{1}{R_p} = \frac{R}{R^2 + X_L^2}$$

$$B_L = \frac{1}{X_p} = \frac{X_L}{R^2 + X_L^2}$$

Some illustrative problems will demonstrate the theory.

---

sample problem

Given the circuit shown with $\omega = 10^6$ rad/sec, determine the equivalent series circuit.
$R = 50\ \Omega$
$C = 0.01\ \mu f$

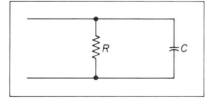

solution

step 1:   Calculate X_c.

$$X_c = \frac{1}{\omega C} = \frac{1}{10^6 \times 10^{-8}}$$

$$X_c = 100\ \Omega$$

step 2:   Calculate R_0.

$$R_0 = \frac{RX_c^2}{R^2 + X_c^2} = \frac{50(100)^2}{2500 + 10,000}$$

$$R_0 = 40 \ \Omega$$

step 3:   Calculate X_0.

$$X_0 = \frac{R^2 X_c}{R^2 + X_c^2} = \frac{(50)^2 100}{12,500}$$

$$X_0 = 20 \ \Omega$$

step 4:   Calculate C_0.

$$C_0 = \frac{1}{\omega X_0} = \frac{1}{10^6(20)}$$

$$C_0 = 0.05 \ \mu F$$

step 5:   Construct the equivalent circuit.

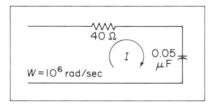

$$Z = 40 - j20$$

sample problem

Given the circuit shown with $\omega = 10^6$ rad/sec, determine the equivalent series circuit.

$R = 200 \ \Omega$

$L = 500 \ \mu H$

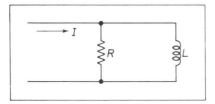

solution

step 1: Calculate X_L.

$$X_L = \omega L = 10^6 \times 500 \times 10^{-6} = 500 \ \Omega$$

step 2: Calculate R_0.

$$R_0 = \frac{R X_L^2}{R^2 + X_L^2} = \frac{200(25 \times 10^4)}{(4 + 25)10^4}$$

$$R_0 = \tfrac{5000}{29} = 172 \ \Omega$$

step 3: Calculate X_0.

$$X_0 = \frac{R^2 X_L}{R^2 + X_L^2} = \frac{4(500)10^4}{29(10^4)}$$

$$X_0 = 69 \ \Omega$$

step 4: Calculate L_0.

$$L_0 = \frac{X_0}{\omega} = \frac{69}{10^6}$$

$$L_0 = 69 \ \mu H$$

step 5: Construct the equivalent circuit.

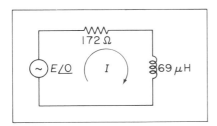

$$Z = 172 + j69$$

sample problem

Given the circuit shown with $\omega = 10^6$ rad/sec, determine the equivalent parallel circuit.

$R = 50 \ \Omega$

$C = 0.01 \ \mu F$

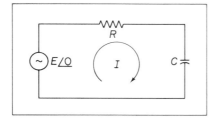

solution

step 1: Calculate X_c.

$$X_c = \frac{1}{\omega C} = \frac{1}{10^6 \times 10^{-8}} = 100 \text{ ohms}$$

step 2: Calculate Y.

$$G = \frac{R}{R^2 + X_c^2} = \frac{50}{2500 + 10,000}$$

$$G = 4 \times 10^{-3}$$

$$B = \frac{X_c}{12,500} = \frac{100}{12,500}$$

$$B = 8 \times 10^{-3}$$

$$Y = (4 - j8)10^{-3} \text{ mhos}$$

In terms of resistance and capacitance, the equivalent circuit becomes

$$R = \frac{1}{G} = \frac{1}{4 \times 10^{-3}} = 250 \text{ } \Omega$$

$$B = \frac{1}{X_c} = \omega C$$

$$C = \frac{B}{\omega} = \frac{8 \times 10^{-3}}{10^6} = 800 \text{ pF}$$

Thus,

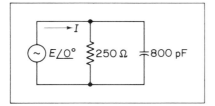

sample problem

Given the circuit shown with $\omega = 10^6$ rad/sec, find the equivalent parallel circuit.

$R = 200 \; \Omega$

$L = 500 \; \mu\text{H}$

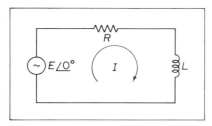

solution

step 1:   Calculate X_L.

$$X_L = \omega L = 10^6 \times 500 \times 10^{-6} = 500 \; \Omega$$

step 2:   Calculate Y.

$$G = \frac{R}{R^2 + X_L^2} = \frac{200}{(4 + 25)10^4}$$

$$G = 6.9 \times 10^{-4}$$

$$B_L = \frac{X_L}{R^2 + X_L^2} = \frac{500}{29 \times 10^4}$$

$$B_L = 17.25 \times 10^{-4}$$

$$Y = (6.9 - j17.25)10^{-4}$$

step 3:   The equivalent circuit is

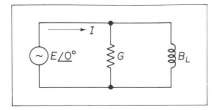

Converting back from admittance to resistance and reactance parameters results in

$$R_p = \frac{1}{G} = \frac{1}{6.9 \times 10^{-4}}$$

$$R_p = 1450\ \Omega$$

$$X = \frac{1}{B_L} = \frac{1}{17.25 \times 10^{-4}}$$

$$X = 580\ \Omega$$

$$L = \frac{X}{\omega} = \frac{580}{10^6}$$

$$L = 580\ \mu\text{H}$$

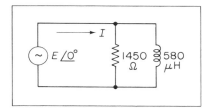

parallel RLC circuits

Three elements are connected in parallel across a common source, as shown in Fig. 14–12. The branch current through the resistor R is in phase with the applied voltage; the branch current through the inductor L lags the applied voltage by 90°, whereas the branch current through the capacitor leads the applied voltage by 90°.

For simplicity of analysis, the admittance technique will be used. Thus,

$$Y_T = Y_1 + Y_2 + Y_3$$

where

$$Y_1 = \frac{1}{R}$$

$$Y_2 = \frac{1}{+jX_L}$$

$$Y_3 = \frac{1}{+jX_c}$$

Then

$$Y_T = G + j(B_c - B_L)$$

or

$$Y_T = \sqrt{G^2 + (B_c - B_L)^2} \; \underline{/\theta}$$

where

$$\theta = \tan^{-1} \frac{B_c - B_L}{G}$$

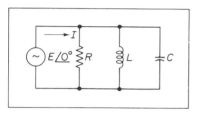

figure 14–12
parallel RLC circuit

The impedance of the circuit can be expressed by

$$Z_T = \frac{Z_1 Z_2 Z_3}{Z_1 Z_2 + Z_1 Z_3 + Z_2 Z_3}$$

where

$$Z_1 = R$$

$$Z_2 = +jX_L$$

$$Z_3 = -jX_c$$

Substituting and further simplification result in

$$Z_T = \frac{RX_L X_c}{X_L X_c + jR(X_L - X_c)}$$

An illustrative problem will demonstrate the theory.

sample problem

Find the total impedance of the following circuit with $\omega = 10^6$ rad/sec.
$R = 200\ \Omega$
$L = 500\ \mu\text{H}$
$C = 0.01\ \mu\text{F}$

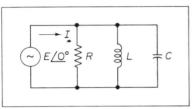

solution

step 1: Calculate Y_1.

$$Y_1 = \frac{1}{R} = \frac{1}{200} = 5 \times 10^{-3}$$

step 2: Calculate Y_2.

$$Y_2 = \frac{1}{+jX_L} = \frac{1}{+j10^6 \times 500 \times 10^{-6}}$$
$$Y_2 = -j2 \times 10^{-3}$$

step 3: Calculate Y_3.

$$Y_3 = \frac{1}{-jX_c} = +j10^6 \times 10^{-8} = 10 \times 10^{-3}$$

step 4: Calculate Y_T.

$$Y_T = Y_1 + Y_2 + Y_3 = [5 + j(10 - 2)]10^{-3}$$
$$Y_T = 9.45 \times 10^{-3}\underline{/58^\circ}$$

step 5: Calculate Z_T.

$$Z_T = \frac{1}{Y_T} = \frac{1}{9.45 \times 10^{-3}\underline{/58^\circ}}$$
$$Z_T = 106\underline{/-58^\circ}$$

step 6: Use the impedance formulas to check solution,

where $R = 200\ \Omega$

$X_L = 500\ \Omega$

$X_c = 100\ \Omega$

$$Z_T = \frac{RX_LX_c}{X_LX_c+jR(X_L-X_c)}$$

$$Z_T = \frac{10 \times 10^6}{(5+j8)10^4}$$

$$Z_T = \frac{1000}{9.45\underline{/58^\circ}} = 106\underline{/-58^\circ}$$

step 7: The equivalent series circuit is

$$Z_T = 106\underline{/-58^\circ} = 56.2-j90$$

$$C = \frac{1}{\omega X_c} = \frac{1}{10^6(90)} = 0.0111\ \mu F$$

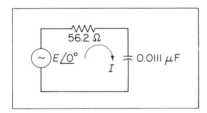

A parallel RLC circuit that is quite useful in electrical technology and engineering is shown in Fig. 14–13.

The impedance of the RL branch is $Z_L = R+jX_L$. The impedance of the capacitive branch is $Z_c = -jX_c$. The impedance of the circuit can be determined by the product over sum rule, namely,

$$Z = \frac{Z_LZ_c}{Z_L+Z_c} = \frac{(R+jX_L)(-jX_c)}{R+j(X_L-X_c)}$$

The circuit can also be analyzed by the admittance method. Thus,

$$Y_1 = \frac{1}{Z_L} = \frac{1}{R+jX_L}$$

$$Y_2 = \frac{1}{Z_c} = \frac{1}{-jX_c}$$

$$Y_T = Y_1+Y_2 = \frac{1}{R+jX_L} + \frac{1}{-jX_c}$$

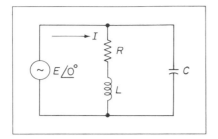

figure 14–13
parallel *RLC* circuit

$$Y_T = \underbrace{\left(\frac{R}{R^2 + X_L^2}\right)}_{G} - j\underbrace{\left(\frac{X_L}{R^2 + X_L^2}\right)}_{B_L} + j\underbrace{\left(\frac{1}{X_c}\right)}_{B_c}$$

$$Y_T = G + j(B_c - B_L)$$

$$Y_T = \sqrt{G^2 + (B_c - B_L)^2} \ \underline{/\theta}$$

$$\theta = \tan^{-1}\frac{(B_c - B_L)}{G}$$

An illustrative problem will demonstrate the theory.

sample problem

Given the circuit shown with $\omega = 10^6$ rad/sec, find: (a) Z_T, (b) I.
$R = 400 \ \Omega$ $C = 2500 \ \text{pF}$
$L = 200 \ \mu\text{H}$ $E = 100 \ \text{v}$

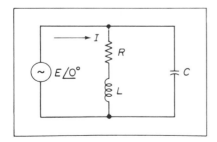

solution

step 1: Calculate X_L and X_c.

$$X_L = \omega L = 10^6(200)10^{-6} = 200 \ \Omega$$

$$X_c = \frac{1}{\omega C} = \frac{1}{10^6 \times 2500 \times 10^{-12}}$$

$$X_c = 400 \ \Omega$$

step 2: Calculate Z_T.

$$Z_T = \frac{(R+jX_L)(-jX_c)}{R+j(X_L-X_c)}$$

$$Z_T = \frac{(400+j200)(-j400)}{400-j200}$$

step 3: Simplify Z_T.

$$Z_T = 320-j240$$
$$Z_T = 400\underline{/-36.9°}$$

step 4: Calculate I.

$$I = \frac{E}{Z_T} = \frac{100\underline{/0°}}{400\underline{/-36.9°}}$$

$$I = 0.25\underline{/36.9°} \ A$$

series parallel ac circuits

Consider the series parallel circuit shown in Fig. 14–14. The problem in this circuit is to determine the value of the line current I. The first step in the solution is to convert the parallel combination of L and C to an equivalent series element. Thus, the impedance of the parallel combination is

$$Z_p = \frac{(j60)(-j40)}{j(60-40)}$$

$$Z_p = -j120 \ \Omega$$

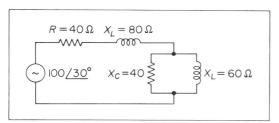

figure 14–14
series parallel *ac* circuit

The next step is to calculate the total impedance presented to the source. Since the parallel combination appears as a single capacitor having a capacitive reactance of 120 ohms, the total series impedance is

$$Z_T = 40 + j(80 - 120) = 40 - j40$$
$$Z_T = 56.56 \underline{/-45°} \ \Omega$$

The evaluation of the line current is the next step. Thus,

$$I = \frac{E}{Z_T} = \frac{100 \underline{/30°}}{56.56 \underline{/-45°}} = 1.77 \underline{/75°} \ A$$

Note the step-by-step method used to solve a series parallel circuit problem. These steps were

step 1

Convert the parallel circuit to an equivalent series circuit.

step 2

Sum up all the series impedances algebraically. This value defines Z_T.

step 3

Divide Z_T into the source voltage to determine the line current.

An illustrative problem will demonstrate the theory.

---

sample problem

Given the circuit shown, determine (a) Z_T, (b) I, (c) I_R.

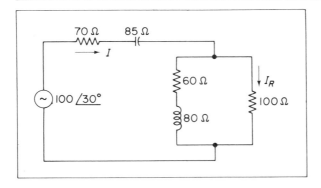

solution

step 1:   Convert the parallel combination to an equivalent series circuit.

$$Z_p = \frac{(60+j80)100}{160+j80} = \frac{2000(3+j4)}{80(2+j1)}$$

$$Z_p = 50+j25$$

step 2:   Calculate Z_T.

$$Z_T = 70-j85+50+j25 = 120-j60$$
$$Z_T = 135\underline{/-26.5°}\ \Omega$$

step 3:   Calculate I.

$$I = \frac{E}{Z_T} = \frac{100\underline{/30°}}{135\underline{/-26.5°}} = 0.74\underline{/56.5°}\ A$$

step 4:   Calculate I_R, using the current divider rule.

$$I_R = 0.74\underline{/56.5°}\ \frac{60+j80}{160+j80}$$

$$I_R = 0.74\underline{/56.5°}\ \frac{100\underline{/53.1°}}{179\underline{/26.5°}} = 0.414\underline{/83.1°}\ A$$

step 5:   The phasor diagram shows the relative positions of the currents.

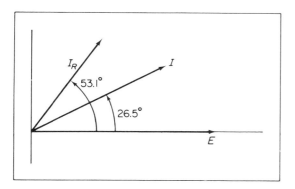

problems

1. A 4-kilohm resistor and a 0.002-μF capacitor are connected in parallel across a 240-V at a frequency of $(10^5/\pi)$ source. Determine the circuit impedance and the line current.

2. A 600-ohm resistor is placed in parallel with a 10-pF capacitor across a 100-V at a frequency of $(10^8/\pi)$ voltage source. Find: (a) Z_T, (b) I.

3. A 3-kilohm resistor is in parallel with a 300-pF capacitor. The applied voltage across the combination is a 300-V at a frequency of $(10^6/3.6\pi)$ voltage source. Find: (a) Z_T, (b) I.

4. A 1-kilohm resistor is placed in parallel with a 200-pF capacitor and a 200-V at a frequency of $(10^6/0.8\pi)$ voltage source. Find: (a) Z_T, (b) I.

5. The magnitude of an impedance at 15 kHz is 5 kilohms. When the frequency is increased to 60 kHz, the impedance drops to 2 kilohms. Determine the parallel combination of R and C required to produce these results.

6. The impedance of a circuit of a resistor in parallel with a capacitor is 6000 ohms at an angle of $65°$. If the frequency is $(10^6/\pi)$, determine R and C.

7. What value of capacitance must be connected in parallel with a 0.1-kilohm resistor to form a total admittance of 30 millimhos at $(10^4/\pi)$?

8. A 120-pF capacitor is connected across a resistor. The voltage applied across the circuit is 100 V at a frequency of $(1.5(10^6)/\pi)$. What is the value of R if the circuit impedance is 2 kilohms?

9. An inductor of 25 mH is in parallel with a resistor of 40 ohms. The voltage applied across the circuit is 100 V at a frequency of $(10^3/2\pi)$. Determine: (a) Z_T, (b) I.

10. A 50-mH coil is in parallel with a 300-ohm resistor. The voltage applied is 50 V at a frequency of $(9 \times 10^3/2\pi)$. Find: (a) Z_T, (b) I.

11. A 4-kilohm resistor is in parallel with a 5-H coil. The voltage applied is 50 V at a frequency of $(10^3/\pi)$. Find: (a) Z_T, (b) I.

12. What is the magnitude of the impedance when a 400-μH coil is placed in parallel with a 600-ohm resistance? The input source voltage is 100 V at a frequency of $(10^6/\pi)$. Find I.

13. The admittance of a parallel RL circuit is $0.04 - j0.02$ mmhos at an input frequency of $(2 \times 10^6/\pi)$. Find R and L.

14. What value of resistance must be connected in parallel with an inductor of 2 H to produce a magnitude of impedance of 2000 ohms at a frequency of $(10^3/\pi)$ Hz?

15. The impedance of a parallel combination of R and L is 4000 ohms at an angle of 65°. The input frequency is $(10^6/\pi)$. Determine the values of R and L.

16. Given the following circuit, find: (a) Z_T, (b) I.

$\omega = 10^5$ rad/sec $C = 0.004$ μF
$R = 10$ kΩ $E = 100$ v
$L = 20$ mH

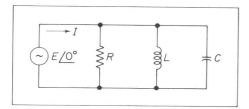

17. Given the following circuit, find: (a) Z_T, (b) I.

$\omega = 4 \times 10^3$ $C = 2$ μF
$R = 200$ Ω $E = 100$ v
$L = 40$ mH

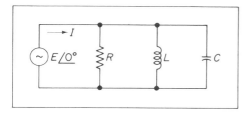

18. A resistor of 10 kilohms is connected in parallel with a capacitive reactance of 5 kilohms. Determine the equivalent series circuit.

19. A resistor of 20 kilohms is connected in parallel with a capacitive reactance of 15 kilohms. Determine the equivalent series circuit.

20. A resistor of 1 kilohm is connected in parallel with an inductive reactance of 3 kilohms. Determine the equivalent series circuit.

21. A resistor of 2.5 kilohms is connected in parallel with an inductive reactance of 5 kilohms. Determine the equivalent series circuit.

22. A series circuit consists of a 3-kilohm resistor and a capacitive reactance of 4 kilohms. Determine the equivalent parallel circuit.

23. A series circuit consists of a 10-kilohm resistor and a capacitive reactance of 6 kilohms. Determine the equivalent parallel circuit.

24. A series circuit consists of a 15-kilohm resistor and an inductive reactance of 10 kilohms. Determine the equivalent parallel circuit.

25. A series circuit consists of a 4-kilhom resistor and an inductive reactance of 9 kilohms. Determine the equivalent parallel circuit.

26. Given the circuit shown with $\omega = 10^3$ rad/sec, find: (a) Z_T, (b) I.

$R = 200\ \Omega$ $\quad$ $C = 2.5\ \mu F$

$L = 500\ mH$ $\quad$ $E = 100\underline{/0°}$

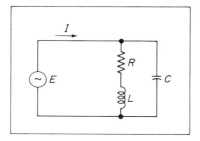

27. Given the circuit shown, find: (a) Z_T, (b) I.

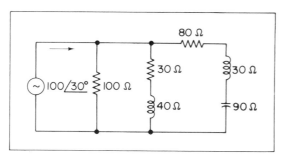

28. Given the circuit shown, find: (a) Z_T, (b) I.

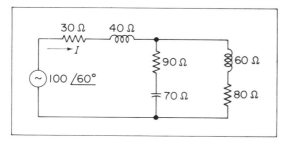

29. Given the circuit shown, find: (a) Z_T, (b) I.

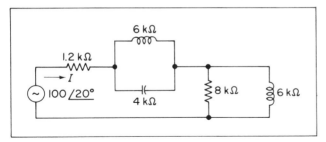

30. Given the circuit shown, find: (a) Z_T, (b) I.

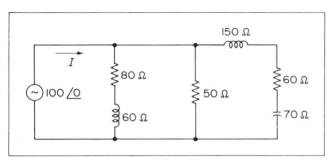

15

resonance

series resonant circuit

The motion of a pendulum in a clock is a mechanical form of a condition called resonance. Note that the amplitude of swing is constant as the pendulum traces out a sinusoidal waveform in its swing.

A similar action can occur in an electrical system comprised of inductors and capacitors. When resonance occurs, the energy stored at any instant of time by one reactive element within the system is precisely the same as the energy given up by another reactive element within the system. The condition of resonance or oscillation implies that the system is self-sustaining. The total power absorbed by the system is equal to the power dissipated across the resistive elements.

Consider the series *RLC* circuit shown in Fig. 15–1.

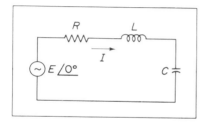

figure 15–1
series *RLC* circuit

The general formula for the impedance of a series *RLC* circuit is

$$Z = R + j(X_L - X_c)$$

where

$$X_L = \omega L \quad \text{and} \quad X_c = \frac{1}{\omega C}$$

It is evident that X_L varies linearly with frequency, having a slope equal to L. On the other hand, X_c varies inversely with frequency, tracing out a hyperbolic path. Both relationships are shown in Fig. 15–2. As frequency is varied, there will be a frequency where the reactance of the inductor

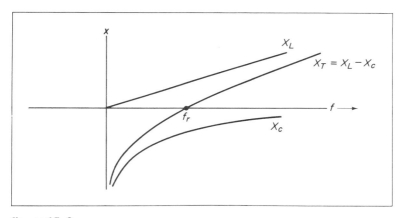

figure 15–2
reactance vs. frequency

will be exactly equal to the reactance of the capacitor. This particular frequency (f_r) is called the *resonant frequency*.

A definition of series resonance, therefore, is one in which the total reactance is equal to zero and the impedance of the circuit appears as a pure resistance.

Using the specified definition of resonance, we can determine the frequency at which resonance occurs by setting the reactive term equal to zero. Thus,

$$X_L = X_c \quad \text{at } f = f_r$$

$$\omega_r L = \frac{1}{\omega_r C}$$

Since $\omega_r = 2\pi f_r$, then substituting and simplifying yields

$$f_r = \frac{1}{2\pi\sqrt{LC}} \text{ Hz}$$ (15–1)

The impedance of a circuit varies as a function of frequency, as shown in Fig. 15–3. Note that at resonance $Z = R + j0$. The current in the circuit at resonance is

$$I_0 = \frac{E}{R}$$

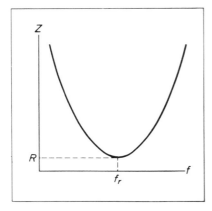

figure 15–3
impedance vs. frequency

The voltage drop across the resistor at resonance is equal to the applied voltage, since the voltage drop is $E_R = I_0 R = E$. At this frequency, the voltage drop across the inductor is equal to and 180° out of phase with the voltage drop across the capacitor. At frequencies below resonance, the capacitive reactance is greater than the inductive reactance and the circuit is capacitive in nature. At frequencies above resonance, the inductive reactance predominates and the circuit is inductive in nature. At the resonant frequency, the circuit is purely resistive, since the net reactance is zero.

The ratio between the peak power stored in either the capacitor or inductor at resonance with respect to the average power that is absorbed by the resistive element is called the *quality factor* of that resonant circuit. The letter symbol for the quality factor is Q.

$$Q = \frac{P_x}{P_R}$$

In general, the resistance in the circuit is usually the coil resistance, and the discussion usually concerns the Q *of the coil.* The Q of the coil at resonance is designated as Q_0 and is equal to

$$Q_0 = \frac{I_0^2 X_L}{I_0^2 R} = \frac{\omega_r L}{R}$$

(15–2)

The term Q_0, therefore, is used as a figure of merit to classify and compare different coils. It is evident that Q_0 is a dimensionless unit, since it is defined as the ratio of powers or the ratio of ohms. Using field terminology, we say that a coil having a Q_0 less than 10 is a low Q coil, whereas the Q_0 greater than 10 is a high Q coil.

An illustrative problem will demonstrate the theory.

sample problem

A series resonant circuit consists of a 100-μH coil and a 100-pF capacitor. The circuit resistance is 50 ohms. Determine: (a) f_r, (b) Q_0.

solution

step 1: Solve for f_r.

$$f_r = \frac{1}{2\pi\sqrt{LC}} = \frac{0.159}{10^{-7}}$$

$$f_r = 1.59 \text{ MHz}$$

step 2: Calculate Q_0.

$$Q_0 = \frac{2\pi f_r L}{R} = \frac{2\pi \times 1.59 \times 10^6 \times 100 \times 10^{-6}}{50}$$

$$Q_0 = 20$$

sharpness of resonance

The sharpness of resonance is defined as the ratio of the band width of the circuit to the resonant frequency. The band width of a circuit is arbitrarily defined as the range of frequencies wherein the current is equal to or greater than a specified decimal value of the maximum current. In this text, the frequency band width will be specified at 0.707 of the maximum current. A graph of current versus frequency indicating band width measurement (Δf) is shown in Fig. 15–4. The mathematical statement for the sharpness of resonance is

$$\text{Sharpness of resonance} = \frac{f_2 - f_1}{f_r} = \frac{\Delta f}{f_r}$$

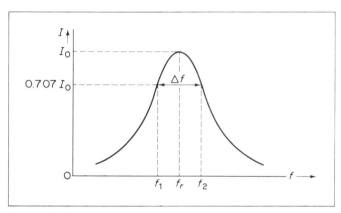

figure 15–4
current vs. frequency in a series *RLC* circuit

analysis

The magnitude of current at any frequency can be expressed by

$$I = \frac{E}{\sqrt{R^2 + (X_L - X_c)^2}}$$

As frequency is varied below resonance, the magnitude of reactance will equal the resistance of the circuit at some frequency symbolized by f_1. Since the circuit is capacitive below resonance, the mathematical relationship that must be used is

$$X_L - X_c = -R \quad \text{at } f = f_1$$

or

$$\frac{1}{\omega_1 C} - \omega_1 L = +R$$

At a frequency above resonance, again the magnitude of the reactance will equal the resistance in the circuit. This frequency is symbolized by f_2.

The circuit at this frequency is inductive and the mathematical relationship is

$$X_L - X_c = +R \quad \text{at } f = f_2$$

or

$$\omega_2 L - \frac{1}{\omega_2 C} = R$$

Solving the two equations simultaneously yields

$$f_2 - f_1 = \frac{R}{2\pi L}$$

Dividing both sides by f_r results in the following relationship for the sharpness of resonance.

$$\boxed{\frac{\Delta f}{f_r} = \frac{1}{Q_0}} \tag{15-3}$$

From this equation, it is evident that as the value of Q_0 increases, the band width decreases. As the band width decreases, the selectivity of the circuit increases.

maximum voltage considerations

At resonance, the current in the circuit is maximum and equal to

$$I_0 = \frac{E}{R}$$

The magnitude of voltage across the inductor is

$$E_L = I_0 X_L = \frac{E}{R} X_L = Q_0 E$$

The magnitude of voltage across the capacitor is exactly the same, since $X_L = X_c$. In a series RLC circuit, having an applied voltage of 10 volts and a circuit Q of 50, the voltage across either the inductor or the capacitor is 500 volts at resonance.

The frequency at which maximum voltage exists across the capacitor can be determined by use of differential calculus.

The magnitude of current at any frequency is

$$I = \frac{E}{\sqrt{R^2 + \left(\omega L - \dfrac{1}{\omega C}\right)^2}}$$

The magnitude of voltage across the capacitor is

$$E_c = \frac{E}{\sqrt{R^2 \omega^2 C^2 + (\omega^2 LC - 1)^2}}$$

A graph of E_c with respect to ω for different values of Q is shown in Fig. 15–5.

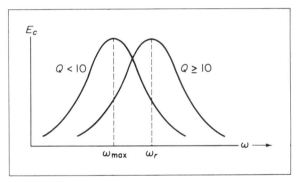

figure 15–5
E_c vs. ω for varying Q

The frequency at which maximum voltage occurs across the capacitor is given by the mathematical relationship

$$f_{max} = \frac{1}{2\pi\sqrt{LC}} \times \sqrt{1 - \frac{1}{2Q^2}} \tag{15–4}$$

Note that the voltage maximum occurs at a frequency below resonance when $Q < 10$. If $Q \geq 10$, then the voltage maximum occurs at approximately the resonant frequency.

The frequency at which maximum voltage exists across the inductor is evaluated in a similar manner by graphing the magnitude of E_L with respect to ω and varying Q accordingly. The mathematical resultant is

$$f_{max} = \frac{1}{2\pi\sqrt{LC}} \times \frac{1}{\sqrt{1 - \frac{1}{2Q^2}}}$$

(15–5)

A graph of the voltages versus frequency is shown in Fig. 15–6.

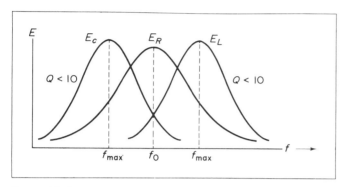

figure 15–6
voltage vs. frequency variations

parallel resonance

A coil and a capacitor are connected in parallel, as shown in Fig. 15–7. The quadrature component of the current in the inductive branch must be exactly equal to and 180° out of phase with the current in the capacitive branch, at resonance. A definition of a parallel resonant circuit is one in which the total susceptance is zero. The impedance of the circuit appears as a pure resistance.

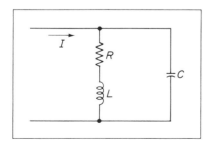

figure 15–7
parallel *RLC* circuit

The total admittance of the circuit is given by

$$Y_{in} = Y_L + Y_c$$

where

$$Y_L = \frac{1}{R+jX_L} = \frac{R}{R^2+X_L^2} -j\frac{X_L}{R^2+X_L^2}$$

$$Y_c = \frac{1}{-jX_c}$$

Thus,

$$Y_{in} = \frac{R}{R^2+X_L^2} +j\left(\frac{1}{X_c} - \frac{X_L}{R^2+X_L^2}\right)$$

This equation is in the form of

$$Y_{in} = G+jB$$

where

$$B = B_c - B_L$$

The resonant frequency f_{ar} at which the parallel circuit is resonant can be determined as follows.

$$B_L = B_c \quad \text{at } f = f_{ar}$$

and

$$\frac{X_L}{R^2+X_L^2} = \frac{1}{X_c}$$

$$\frac{\omega L}{R^2+\omega^2L^2} = \omega C$$

$$\omega_{ar} = \sqrt{\frac{1}{LC} - \frac{R^2C}{L^2C}}$$

or

$$\boxed{f_{ar} = f_r \sqrt{1 - \frac{1}{Q_0^2}}} \qquad (15\text{–}6)$$

where

$$Q_0 = \frac{\omega_r L}{R}$$

$$f_r = \frac{1}{2\pi \sqrt{LC}}$$

The impedance of the circuit at resonance symbolized by R_{ar} is equal to the reciprocal of the conductance. Thus,

$$R_{ar} = \frac{R^2 + X_L^2}{R} = R(1 + Q_0^2) \tag{15-7}$$

If the Q_0 of the coil is equal to or greater than 10, then

$$R_{ar} = Q_0^2 R = \frac{X_L^2}{R} = \frac{X_c^2}{R} = \frac{L}{RC} \tag{15-8}$$

The typical susceptance curves are shown in Fig. 15–8.

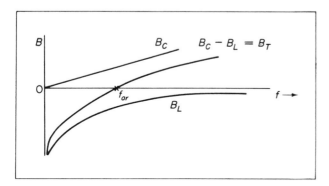

figure 15–8
susceptance vs. frequency

The reactance curves are obtained by taking the reciprocal of the susceptance curves B_T, which yields the resultant curves shown in Fig. 15–9.

It is evident that at frequencies below resonance the circuit is inductive, whereas at frequencies above resonance the circuit is capacitive. At the antiresonant frequency, both ends of the reactance curve go to infinity, the inductive branch in a positive direction and the capacitive branch in a

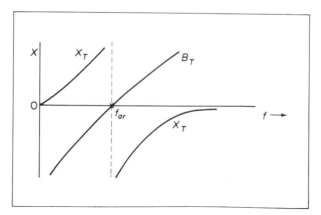

figure 15–9
reactance vs. frequency

negative direction. The source current is a minimum in the circuit at anti-resonance and equal to zero if the branch resistances are nonexistent. A circuit of this type is used for wave trap operation.

parallel resonance with resistance in both branches

A parallel circuit containing resistance in both branches is shown in Fig. 15–10. The criterion for resonance of the circuit is unity power factor. Then the susceptance of the inductor is equal to the susceptance of the capacitor. Thus,

$$B_L = B_c \quad \text{at } f = f_{ar}$$

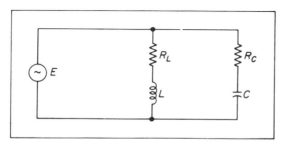

figure 15–10
parallel $RLRC$ circuit

The frequency at which parallel resonance occurs is determined by the following procedure.

$$\frac{X_L}{R_L^2 + X_L^2} = \frac{X_c}{R_c^2 + X_c^2}$$

Simplifying and solving results in

$$\frac{\omega_{ar} L}{R_L^2 + \omega_{ar}^2 L^2} = \frac{\omega_{ar} C}{R_c^2 \omega_{ar}^2 C^2 + 1}$$

The solution for f_{ar} is

$$f_{ar} = f_r \left(\frac{R_L^2 C - L}{R_c^2 C - L} \right)^{1/2} \tag{15-9}$$

The conductance of the circuit at resonance is

$$G_{ar} = \frac{R_L}{R_L^2 + X_L^2} + \frac{R_c}{R_c^2 + X_c^2}$$

Note that since

$$\left(\frac{1}{R_c^2 + X_c^2} \right) = \frac{X_L / X_c}{R_L^2 + X_L^2}$$

then

$$G_{ar} = \frac{R_L + \dfrac{R_c X_L}{X_c}}{R_L^2 + X_L^2}$$

$$G_{ar} = \frac{R_L X_c + R_c X_L}{X_c (R_L^2 + X_L^2)}$$

The impedance of the circuit at resonance is the reciprocal of the conductance. Thus,

$$R_{ar} = \frac{X_c (R_L^2 + X_L^2)}{R_L X_c + R_c X_L}$$

If the circuit is a high Q circuit, that is, Q_0 is equal to or greater than

10, and the magnitude of X_L is approximately equal to the magnitude of X_c, then

$$R_{ar} = \frac{L}{C(R_c + R_L)} \qquad (15\text{-}10)$$

Consider the parallel resonance circuit shown in Fig. 15–11. At parallel resonance, the source current is given by

$$I_{ar} = \frac{E}{R_{ar}}$$

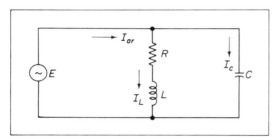

figure 15–11
parallel RLC circuit

The current in the capacitive branch can be evaluated by the current divider rule. Thus,

$$I_c = I_{ar} \left(\frac{R + jX_L}{R + j(X_L - X_c)} \right)$$

If the following two assumptions are made, the results are given by Eq. (15–11).

1. $Q_0 \geq 10$

2. $X_L = X_c$

$$I_c = Q_0 I_{ar} \qquad (15\text{-}11)$$

The results of this analysis may be exlapined in the following manner. If the source current is 2 amperes and the Q_0 of the coil at resonance is 100, then the branch currents that flow are 200 amperes, respectively. Note that at resonance $I_L = I_c$, provided that the two assumptions are met.

sharpness of resonance

The sharpness of resonance for a series circuit is equal to

$$\frac{\Delta f}{f_r} = \frac{1}{Q_0}$$

Consider the general circuit shown in Fig. 15–12. Thevenizing the circuit will yield a simple series equivalent circuit, and the equation for a simple series circuit may be used.

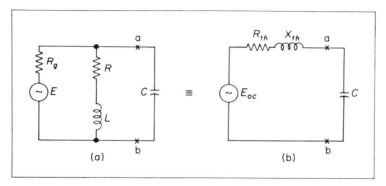

figure 15–12
(a) parallel *RLC* circuit; (b) Thevenin's equivalent

analysis

$$Z_{th} = \frac{R_g(R+jX_L)}{R+R_g+jX_L}$$

The assumption will be made that $R_g \gg R$. Then, rationalizing the expression, we find that the result becomes

$$Z_{th} = \frac{R_g^2 R + R_g X_L^2 + jR_g^2 X_L}{R_g^2 + X_L^2} = R_{th} + jX_{th}$$

The sharpness of resonance is equal to $1/Q_0$ or

$$\frac{f_2-f_1}{f_{ar}} = \frac{R_{th}}{X_{th}} = \frac{R}{X_L} + \frac{X_L}{R_g}$$

$$\boxed{\frac{f_2-f_1}{f_{ar}} = \frac{1}{Q_0} + \frac{X_L}{R_g}}$$

(15–12)

Note that if $R_g = R_{ar}$ for maximum power transfer then the sharpness of resonance is

$$\frac{f}{f_{ar}} = \frac{2}{Q_0}$$

(15–13)

An illustrative problem will demonstrate the theory.

---

sample problem

A 100-pF capacitor is in parallel with a coil that has a resistance of 20 ohms and an inductance of 400 μH. Determine: (a) f_{ar}, (b) R_{ar}, (c) Δf if $R_g = R_{ar}$.

---

solution

step 1: Calculate f_r.

$$f_r = \frac{1}{2\pi\sqrt{LC}} = \frac{1}{2\pi\sqrt{4 \times 10^{-14}}}$$

$$f_r = 795 \text{ kHz}$$

step 2: Calculate f_{ar}.

$$f_{ar} = f_r\sqrt{1 - \frac{R^2C}{L}}$$

$$f_{ar} = 795 \times 10^3\sqrt{1 - 10^{-4}}$$

$$f_{ar} = 795 \text{ kHz}$$

step 3: Calculate Q_0.

$$Q_0 = \frac{\omega_r L}{R} = \frac{2\pi \times 795 \times 10^3 \times 400 \times 10^{-6}}{20}$$

$$Q_0 = 100$$

step 4: Calculate R_{ar}.

$$R_{ar} = \frac{L}{RC} = \frac{400 \times 10^{-6}}{20 \times 10^{-10}}$$

$$R_{ar} = 200 \text{ k}\Omega$$

step 5: Calculate Δf.

$$\frac{\Delta f}{f_{ar}} = \frac{2}{Q_0}$$

$$\Delta f = \frac{2 \times 795 \times 10^3}{100} = 15.9 \text{ kHz}$$

---

sample problem

Design a parallel *RLC* circuit to match a generator of 50 kΩ and have a band width from 975 kHz to 1025 kHz. The antiresonant frequency is 1000 kHz.

---

solution

step 1: Calculate Q_0 of the coil.

$$\frac{\Delta f}{f_{ar}} = \frac{2}{Q_0}$$

$$Q_0 = \frac{2 f_{ar}}{\Delta f} = \frac{2 \times 10^6}{50 \times 10^3}$$

$$Q_0 = 40$$

step 2: Calculate X_L and L.

$$R_{ar} = Q_0 X_L = 40 \times X_L = 50 \times 10^3$$

$$X_L = 1250 \ \Omega$$

$$L = \frac{X_L}{\omega_{ar}} = \frac{1250}{2\pi \times 10^6}$$

$$L = 199 \ \mu\text{H}$$

step 3: Calculate C.

$$X_c = X_L = 1250 \ \Omega$$

$$C = \frac{1}{\omega_{ar} X_c} = \frac{1}{2\pi \times 10^6 \times 1250}$$

$$C = 127 \text{ pF}$$

step 4: Calculate R.

$$R = \frac{X_L}{Q_0} = \frac{1250}{40}$$

$$R = 31.25 \ \Omega$$

step 5:   The circuit that has been designed is

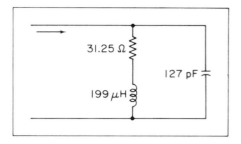

maximum impedance

The antiresonant frequency of a parallel circuit is usually considered as that frequency which makes the circuit a maximum-impedance, minimum-current circuit. The following analysis will determine the frequency where maximum impedance occurs for the circuit shown in Fig. 15–13.

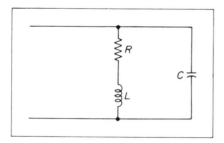

figure 15–13
parallel *RLC* circuit

analysis

The magnitude of impedance is

$$|Z|^2 = \frac{R^2 + \omega^2 L^2}{(1 - \omega^2 LC)^2 + \omega^2 R^2 C^2}$$

Graphing this function with respect to ω yields Fig. 15–14.

The mathematical relationship for the frequency where maximum impedance occurs is given by

$$\omega_{max} = \frac{1}{2\pi\sqrt{LC}}\sqrt{\sqrt{1 + \frac{2R^2 C}{L} - \frac{R^2 C}{L}}} \tag{15–14}$$

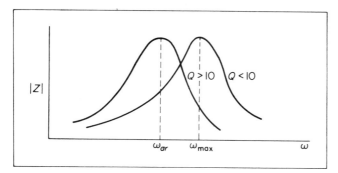

figure 15–14
magnitude vs. ω for different values of Q

In circuits with high Q coils, the frequency that causes the circuit to have maximum impedance becomes equal to the antiresonant frequency.

multiple resonance

In a parallel RLC circuit, the circuit is inductive at frequencies below resonance, whereas at frequencies above resonance, the circuit is capacitive. Consequently, a series element can be inserted to resonate with the parallel RLC circuit at some frequency either above or below the antiresonant frequency. Consider the circuit shown in Fig. 15–15.

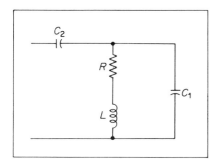

figure 15–15
series parallel circuit

The basic understanding of the circuit operation can be followed by the reactance curves shown in Fig. 15–16. The procedure used to construct these curves is detailed in a step-by-step manner.

step 1

Sketch the reactance curves for the parallel resonant circuit designated as X_p.

step 2

Sketch the reactance curve for C_2 designated as X_2.

step 3

Add the reactance curves X_p and X_2 as shown.

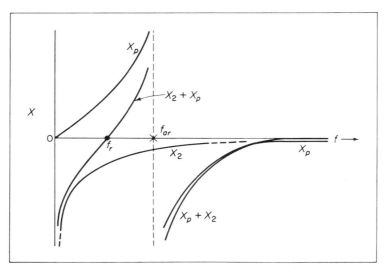

figure 15–16
reactance curves vs. frequency

analysis

The mathematical analysis requires the solution to the parallel circuit first. The next step is to calculate the inductive or capacitive reactance at the series resonant frequency. An illustrative example will demonstrate the theory.

sample problem

A parallel RLC circuit has the following given components:

$$R_L = 90 \ \Omega, \qquad L = 9.54 \text{ mH}, \qquad C = 265 \text{ pF}$$

A reactive element is inserted in series with the parallel circuit to resonate with the parallel network at 200 kHz. Specify the circuit element value and the R_{ar} of the parallel circuit at f_{ar}.

solution

step 1: Calculate f_{ar}.

$$f_{ar} = \frac{1}{2\pi}\sqrt{\frac{1}{LC} - \frac{R_L^2 C}{L^2 C}}$$

$$f_{ar} = 100 \text{ kHz}$$

step 2: Calculate R_{ar}.

$$R_{ar} = \frac{L}{CR_L} = \frac{9.54 \times 10^{-3}}{265 \times 10^{-12} \times 90}$$

$$R_{ar} = 400 \text{ k}\Omega$$

step 3: Calculate Z of the parallel circuit at 200 kHz.

$$Z = \frac{(R_L + jX_L)(-jK_c)}{R_L + j(X_L - X_c)}$$

$$X_L = \omega L = 2\pi \times 200 \times 10^3 \times 9.54 \times 10^{-3}$$
$$X_L = 12,000 \ \Omega$$

$$X_c = \frac{1}{\omega C} = \frac{1}{2\pi \times 200 \times 10^3 \times 265 \times 10^{-12}}$$

$$X_c = 3000 \ \Omega$$

$$Z = \frac{(90 + j12,000)(-j3000)}{90 + j(12,000 - 3000)}$$

$$Z \cong -j4000 \ \Omega$$

step 4: The circuit element required to be put in series with the parallel circuit is an inductor whose reactance is 4000 ohms at the series frequency

$$X_{L_s} = 4000 = \omega L_s$$

$$L_s = \frac{4000}{2\pi \times 200 \times 10^{-3}} = 3.16 \text{ mH}$$

problems

1. A 100-ohm resistor, a 80-mH coil, and a 0.001-μF capacitor are connected in series with a 200-volt source. Find: (a) f_r, (b) Q_0, (c) Δf, (d) $E_{L_{res}}$, (e) $E_{c_{res}}$.

2. A 50-ohm resistor, a 500-μH coil, and a 500-pF capacitor are connected in series with a 100-volt source. Find: (a) f_r, (b) Q_0, (c) Δf, (d) $E_{L_{res}}$, (e) $E_{c_{res}}$.

3. A capacitor C, a 5-ohm resistor, and a 1-mH coil are connected in series with a 200-V source at a frequency of $(50,000/\pi)$. Find: (a) C_r, (b) Q_0, (c) Δf, (d) $E_{L_{res}}$, (e) $E_{c_{res}}$.

4. A 20-ohm resistor, a 40-μH coil and a capacitor C are connected in series with a 100-V source at a frequency of $(5 \times 10^6/\pi)$. Find: (a) C_{res}, (b) Q_0, (c) Δf, (d) $E_{c_{res}}$.

5. A series circuit consists of a 30-ohm resistor, a 40-μH coil, and a capacitor C. If the voltage source is 210 V at a frequency of $(7.5 \times 10^6/\pi)$ find: (a) C for resonance, (b) Δf, (c) $E_{c_{res}}$.

6. A 10-ohm resistor, a 200-pF capacitor, and a coil L are in series with 50 V at a frequency of $(3 \times 10^6/2\pi)$. Find: (a) L for resonance, (b) Δf, (c) $E_{L_{res}}$.

7. A 25-ohm resistor, a 400-pF capacitor and a coil L are in series with 200-V at a frequency of $(4 \times 10^6/\pi)$. Find: (a) L for resonance, (b) Δf, (c) $E_{L_{res}}$.

8. A series circuit consists of a 20-ohm resistor, a 240-pF capacitor, and a coil L. The source voltage is 200 V at a frequency of $(8 \times 10^6/\pi)$. Find: (a) L for resonance, (b) Δf, (c) $E_{L_{res}}$.

9. A series circuit consists of a 20-ohm resistor, a 80-μH coil, and a 500-pF capacitor. The input voltage is 20 volts. Determine: (a) f_r, (b) Δf, (c) f for $E_{c_{max}}$, (d) f for $E_{L_{max}}$, (e) $E_{c_{res}}$, (f) $E_{L_{res}}$.

10. A series circuit consists of a 10-ohm resistor, a 200-μH coil, and a 150-pF capacitor. The input voltage is 40 volts. Determine: (a) f_r, (b) Δf, (c) f_{max} across the coil, (d) $E_{L_{max}}$, (e) f_{max} across the capacitor, (f) $E_{c_{max}}$.

11. Design a series RLC circuit to have a resonant frequency of 1 mHz and a band width of 40 kHz. The coil resistance is 25 ohms.

12. A series RLC circuit has the following given data:
$\Delta f = 20$ kHz, $Q = 100$, $R = 30 \Omega$
Find: (a) f_r, (b) L and C.

13. A 200-pF capacitor is placed in parallel with a coil that has a resistance of 20 ohms and an inductance of 400 μH. Determine: (a) R_{ar}, (b) f_{ar}.

14. A 500-pF capacitor is placed in parallel with a coil that has a resistance of 4 ohms and an inductance of 80 μH. Find: (a) R_{ar}, (b) f_{ar}.

For Problems 15 through 17 the following circuit is given.

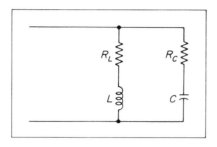

15. In the circuit shown, the given data are
$R_L = 5\ \Omega,$ $L = 250\ \mu$H
$R_c = 2\ \Omega,$ $C = 0.001\ \mu$F
Find: (a) R_{ar}, (b) f_{ar}.

16. In the circuit shown, the given data are
$R_L = 15\ \Omega$ $L = 200\ \mu$H
$R_c = 10\ \Omega$ $C = 0.002\ \mu$F
Find: (a) R_{ar}, (b) f_{ar}.

17. In the circuit shown, the following data are specified.
$R_L = 10\ \Omega,$ $L = 400\ \mu$H
$R_c = 2\ \Omega,$ $C = 0.001\ \mu$F
Find: (a) R_{ar}, (b) f_{ar}.

18. In the resonant circuit shown, the given data are
$R_L = 40\ \Omega,$ $L = 40$ mH, $C = 0.001\ \mu$F, $E = 100$ v
Find: (a) R_{ar}, (b) f_{ar}, (c) I_L, (d) I_c, (e) R_g for maximum power transfer, (f) Δf, (g) f needed for Z_{max}.

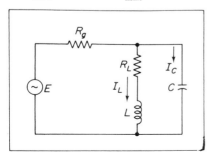

19. In the resonant circuit shown, the generator impedance is equal to the parallel circuit impedance at resonance. The following data are given.

$R_L = 80\ \Omega$, $C = 0.001\ \mu F$, $R_g = 50\ k\Omega$, $E = 100\ v$

Find: (a) L, (b) f_{ar}, (c) Δf, (d) I_L, (e) f required for Z_{max}.

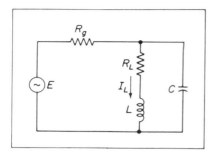

20. Design a parallel RLC circuit to match a generator of 100 kΩ and to have a band width of 40 kHz. The center frequency is 1 MHz.

21. Design a parallel RLC circuit to have an R_{ar} equal to 30 kΩ and a band width of 50 kHz. The generator input impedance is 10 kΩ, and the center frequency of the circuit is 1000 kHz.

16

analysis techniques

introduction

In the solution of typical electrical circuit problems, both time and tedious calculations can be reduced by the application of the proper theorem. Most of these methods for dc circuits were previously discussed in Chapter 5. Consequently, this chapter will consider the requirements and analysis methods for ac circuits only.

loop equations

A conventional method for solving circuits combines mesh currents and Kirchhoff's voltage law applications. Note that for the network shown

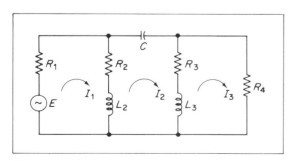

figure 16–1
circuit for loop equations

in Fig. 16–1, there are three independent mesh currents. The arrowhead in all cases is clockwise for simplicity of circuit analysis.

The first step in the analysis is to convert the elements to boxed impedances as shown in Fig. 16–2. The second step applies Kirchhoff's voltage law relationship. The resultant equations for the three loops are

$$E = I_1(Z_1+Z_2)-I_2Z_2+0$$
$$0 = -I_1Z_2+I_2(Z_2+Z_3+Z_5)-I_3Z_3$$
$$0 = 0-I_2Z_3+I_3(Z_3+Z_4)$$

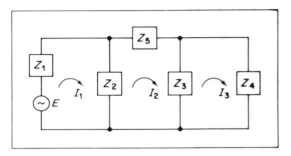

$$Z_1 = R_1$$
$$Z_2 = R_2+jX_2$$
$$Z_3 = R_3+jX_3$$
$$Z_4 = R_4$$
$$Z_5 = -jX_c$$

figure 16–2
converting to boxed impedances

The third step is to evaluate the currents by using determinants. Thus, the solution for I_1 is

$$I_1 = \frac{\begin{vmatrix} E & -Z_2 & 0 \\ 0 & +(Z_2+Z_3+Z_5) & -Z_3 \\ 0 & -Z_3 & +(Z_3+Z_4) \end{vmatrix}}{\begin{vmatrix} (Z_1+Z_2) & -Z_2 & 0 \\ -Z_2 & (Z_2+Z_3+Z_5) & -Z_3 \\ 0 & -Z_3 & (Z_3+Z_4) \end{vmatrix}}$$

As an illustrative problem, assume the following given values for the circuit of Fig. 16–2:

$$Z_1 = 6\,\Omega, \qquad Z_2 = 2+j4, \qquad Z_3 = 3+j6, \qquad Z_4 = 12\,\Omega$$
$$E = 10\underline{/0°}, \qquad Z_5 = -j10$$

Solve for I_1. The denominator value is

$$\Delta = (Z_1+Z_2)(Z_2+Z_3+Z_5)(Z_3+Z_4)-Z_3^2(Z_1+Z_2)-Z_2^2(Z_3+Z_4)$$

Substituting values yields

$$\Delta = (8+j4)(5)(15+j6)-(-27+j36)(8+j4)-(-12+j16)(15+j6)$$
$$\Delta = 1116+j192 = 1132\underline{/9.75°}$$

The numerator of I_1 is

$$N(I_1) = E(Z_2+Z_3+Z_5)(Z_3+Z_4)$$
$$N(I_1) = 10(5)(3)(5+j2) = 750+j300$$
$$N(I_1) = 808\underline{/21.8°}$$

To evaluate I_1, divide the denominator into the numerator. The result is

$$I_1 = \frac{N(I_1)}{\Delta} = \frac{808\underline{/21.8°}}{1132\underline{/9.75°}}$$

$$I_1 = 0.612\underline{/12.05°}\ A$$

---

sample problem

Determine the mesh circuit equations for the given circuit.

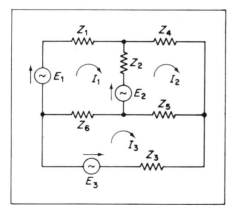

solution

The required three loop equations are

$$E_1-E_2 = I_1(Z_1+Z_2+Z_6)-I_2Z_2-I_3Z_6$$
$$E_2 = -I_1Z_2+I_2(Z_2+Z_4+Z_5)-I_3Z_5$$
$$-E_3 = -I_1Z_6-I_2Z_5+I_3(Z_3+Z_5+Z_6)$$

The remainder of the problem would be solved in the following manner. Given values for the impedances and voltages, the determinant of the denominator would be the first step. Then the second step would solve for the determinant of the numerator depending on the current desired. The ratio of the numerator and denominator specifies the value of the required current. As a student exercise, the following values are given for the problem. Solve for I_1.

$$Z_1 = 4+j6, \qquad Z_4 = 3+j12, \qquad E_1 = 30\underline{/0°}$$
$$Z_2 = 2-j2 \qquad Z_5 = 5-j10 \qquad E_2 = 10\underline{/0°}$$
$$Z_3 = 3+j14 \qquad Z_6 = 2-j4 \qquad E_3 = 40\underline{/0°}$$

nodal analysis

The nodal analysis section in Chapter 5 should be reviewed before one proceeds to the ac applications. This technique is based on Kirchhoff's current law as applied to the nodes of a network. In the review section it was shown that there are $(n-1)$ independent equations (equal to the number of independent nodes minus one) for a given circuit. One node is arbitrarily selected as a reference node, usually ground. The required nodal equations are found by the same procedure used in the review section. An illustrative problem will demonstrate the procedure.

sample problem

Given the circuit shown, find E_a and E_c.

$$I_1 = 6\underline{/0°} \qquad Z_b = +j6$$
$$I_2 = 3\underline{/0°} \qquad Z_c = j2$$
$$Z_a = 4+j0$$

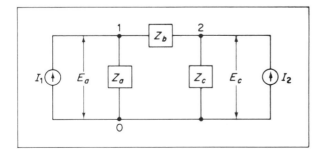

solution

step 1: Write the nodal equations, using Kirchhoff's current laws.

Node 1–0:

$$I_1 = \frac{E_a}{Z_a} + \frac{E_a - E_c}{Z_b}$$

Node 2–0:

$$I_2 = \frac{E_c}{Z_c} + \frac{E_c - E_a}{Z_b}$$

Converting to $Y = 1/Z$ and simplifying yields

$$I_1 = E_a(Y_a + Y_b) - E_c Y_b$$
$$I_2 = -E_a Y_b + E_c(Y_b + Y_c)$$

step 2: Solve for E_a, using determinants.

$$E_a = \frac{I_1(Y_b + Y_c) + I_2 Y_b}{(Y_a + Y_b)(Y_b + Y_c) - Y_b^2}$$

step 3: Evaluate E_a.

$$E_a = \frac{6\left(\dfrac{1}{j6} + \dfrac{1}{-j2}\right) + 3\left(\dfrac{1}{j6}\right)}{\left(\dfrac{1}{4} + \dfrac{1}{j6}\right)\left(\dfrac{1}{j6} + \dfrac{1}{-j2}\right) - \left(\dfrac{1}{j6}\right)^2}$$

$$E_a = \frac{+j1.5}{\frac{1}{12} + j\frac{1}{12}} = 12.72\underline{/45°}\ \text{v}$$

step 4: Solve for E_c, using determinants.

$$E_c = \frac{I_2(Y_a + Y_b) + I_1 Y_b}{\Delta}$$

step 5: Evaluate E_c.

$$E_c = \frac{3\left(\dfrac{1}{4} + \dfrac{1}{j6}\right) + 6\left(\dfrac{1}{j6}\right)}{\frac{1}{12} + j\frac{1}{12}}$$

$$E_c = -(4.5 + j13.5) = 14.25\underline{/251.6°}\ \text{v}$$

sample problem

Write the nodal equations for the given circuit and solve for E_1.

$$I_1 = 10\underline{/0}, \qquad Z_1 = 2-j4, \qquad Z_4 = -j2$$
$$I_2 = 6\underline{/0} \qquad Z_2 = 4-j3 \qquad Z_5 = j4$$
$$Z_3 = +j3 \qquad Z_6 = -j2$$

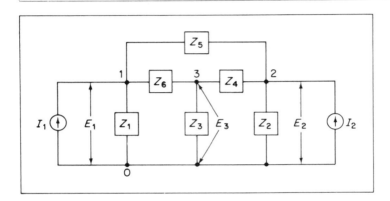

solution

step 1: Write the nodal equations, all nodes with respect to zero.

Node 1:

$$I_1 = E_1 Y_1 + (E_1 - E_3) Y_6 + (E_1 - E_2) Y_5$$

Node 2:

$$I_2 = E_2 Y_2 + (E_2 - E_3) Y_4 + (E_2 - E_1) Y_5$$

Node 3:

$$0 = E_3 Y_3 + (E_3 - E_1) Y_6 + (E_3 - E_2) Y_4$$

step 2: Simplify the node equations.

$$I_1 = E_1(Y_1 + Y_5 + Y_6) - E_2 Y_5 - E_3 Y_6$$
$$I_2 = -E_1 Y_5 + E_2(Y_2 + Y_4 + Y_5) - E_3 Y_4$$
$$0 = -E_1 Y_6 - E_2 Y_4 + E_3(Y_3 + Y_4 + Y_6)$$

step 3: Solve for the denominator Δ.

$$\Delta = \begin{vmatrix} Y_1 + Y_5 + Y_6 & -Y_5 & -Y_6 \\ -Y_5 & Y_2 + Y_4 + Y_5 & -Y_4 \\ -Y_6 & -Y_4 & Y_3 + Y_4 + Y_6 \end{vmatrix}$$

$$\Delta = (Y_1 + Y_5 + Y_6)(Y_2 + Y_4 + Y_5)(Y_3 + Y_4 + Y_6)$$
$$- 2Y_4 Y_5 Y_6 - Y_6^2(Y_2 + Y_4 + Y_5) - Y_4^2(Y_1 + Y_5 + Y_6)$$
$$- Y_5^2(Y_3 + Y_4 + Y_6)$$

step 4: Evaluate the Δ.

$$Y_1 = 0.1 + j0.2, \qquad Y_2 = 0.16 + j0.12, \qquad Y_3 = -j0.33$$
$$Y_4 = +j0.5, \qquad Y_5 = -j0.25 \qquad Y_6 = +j0.5$$

$$\Delta = (0.1 + j0.45)(0.16 + j0.37)(+j0.67) - 2(j0.5)(-j0.25)(j0.5)$$
$$- (j0.5)^2(0.16 + j0.37) - (j0.5)^2(0.1 + j0.45)$$
$$- (-j0.25)^2(j0.67)$$

$$\Delta = -0.073 - j0.1008 - j0.125 + 0.04 + j0.0925$$
$$+ 0.025 + j0.1125 + j0.042$$

$$\Delta = -0.008 + j0.021 = 22.4 \times 10^{-3} \underline{/110.9^\circ}$$

step 5: Solve for $N(E_1)$.

$$N(E_1) = \begin{vmatrix} I_1 & -Y_5 & -Y_6 \\ I_2 & Y_2 + Y_4 + Y_5 & -Y_4 \\ 0 & -Y_4 & Y_3 + Y_4 + Y_6 \end{vmatrix}$$

$$N(E_1) = I_1(Y_2 + Y_4 + Y_5)(Y_3 + Y_4 + Y_6) + I_2 Y_4 Y_6$$
$$- Y_4^2 I_1 + I_2 Y_5(Y_3 + Y_4 + Y_6)$$

$$N(E_1) = 10(0.16 + j0.37)(j0.67) + 6(j0.5)(j0.5)$$
$$- 10(j0.5)^2 + 6(-j0.25)(j0.67)$$

$$N(E_1) = -2.48 + j1.07 - 1.5 + 2.5 + 1$$

$$N(E_1) = -0.48 + j1.07 = 1.165 \underline{/112.7^\circ}$$

step 6: Evaluate E_1.

$$E_1 = \frac{N(E_1)}{\Delta} = \frac{1.165 \underline{/112.7^\circ}}{(22.4 \underline{/110.9^\circ})10^{-3}}$$

$$E_1 = 52 \underline{/1.8^\circ} \text{ v}$$

Note that the actual algebraic sign of the voltage is determined in the resultant solution.

conversion of voltage and current sources

In some circuit problems, it may be convenient to convert a current source to a voltage equivalent or a voltage source to a current equivalent.

Two sources are equivalent with respect to their output terminals if the same current flows through or the same voltage appears across an impedance placed between their output terminals.

Refer to the networks shown in Fig. 16–3. The two sources are equivalent, provided that

$$I_{sc} = \frac{E}{Z_{th}}$$

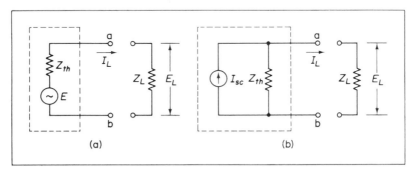

figure 16–3
(a) voltage source; (b) current source

superposition theorem

In general, the theorem states that in any network containing sources and passive, linear, bilateral elements, the current flowing at any point, due to the action of the multiple sources of emf within the network, is equal to the algebraic sum of the currents at this point that would exist if each source were considered independently, all other generators removed and replaced by their internal impedances.

The application of this method to ac circuits requires working with impedances and phasors, rather than resistors and dc voltages.

An illustrative problem will demonstrate the theory.

---

sample problem

Find the current I in the circuit given, using the superposition theorem.

$$Z_1 = 2-j4 \qquad E_1 = 10\underline{/0°}$$
$$Z_2 = 2+j4 \qquad E_2 = 20\underline{/0°}$$
$$Z_3 = 2+j4$$

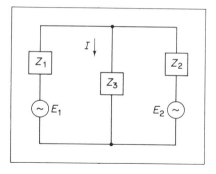

solution

step 1 : Remove E_2 and replace by short circuit. The new circuit is

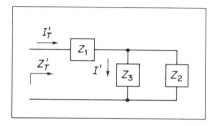

Calculate Z'_T.

$$Z'_T = Z_1 + \frac{Z_2 Z_3}{Z_2 + Z_3}$$

$$Z'_T = 2 - j4 + \frac{(2+j4)(2+j4)}{4+j8}$$

$$Z'_T = 3 - j2$$

step 2 : Calculate I'_T.

$$I'_T = \frac{E_1}{Z'_T} = \frac{10}{3-j2}$$

$$I'_T = 2.307 + j1.54 = 2.78\underline{/33.7°}$$

step 3 : Calculate I'.

$$I' = I'_T \frac{Z_2}{Z_2 + Z_3} = (2.78\underline{/33.7°})\left(\frac{1}{2}\right)$$

$$I' = 1.39\underline{/33.7°} = 1.07 + j0.89$$

step 4: Return to the original circuit. Remove E_1 and replace by short circuit. The new circuit is

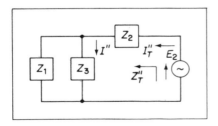

Calculate Z_T''.

$$Z_T'' = Z_2 + \frac{Z_1 Z_3}{Z_1 + Z_3}$$

$$Z_T'' = 2 + j4 + \frac{(2+j4)(2-j4)}{4}$$

$$Z_T'' = 7 + j4$$

step 5: Calculate I_T''.

$$I_T'' = \frac{E_2}{Z_T''} = \frac{20}{7+j4}$$

$$I_T'' = 2.16 - j1.23 = 2.5\underline{/-29.7^\circ}$$

step 6: Calculate I''.

$$I'' = I_T'' \frac{Z_1}{Z_1 + Z_3}$$

$$I'' = (2.5\underline{/-29.7^\circ})(0.5 - j1)$$

$$I'' = 2.8\underline{/-93.2^\circ} = -0.157 - j2.785$$

step 7: Calculate I.

$$I = I' + I''$$

$$I = 1.07 + j0.89 - 0.157 - j2.785$$

$$I = 0.913 - j1.895$$

Thevenin's theorem

The definition of Thevenin's theorem given previously in Chapter 5 is valid for ac circuits, provided that the term *impedance* replaces the word *resistance*. The definition states that any two-terminal network containing voltage and current sources in combination with passive linear bilateral elements can be represented by a series equivalent circuit consisting of a voltage source E_{oc} in series with an impedance Z_{th}.

It is evident that, since the impedance of reactive elements varies with frequency, the theorem is single-frequency applicable.

An example will demonstrate the theory.

---

sample problem

Find the Thevenin's equivalent circuit, using Z_L as load in Fig. 16–4.

$$Z_1 = 3+j2 \qquad Z_3 = 2-j2$$
$$Z_2 = 1-j2 \qquad E = 10\underline{/0°}$$

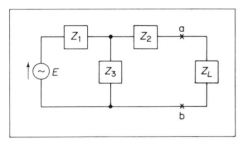

figure 16–4
circuit for Thevenin's theorem

solution

step 1: Determine Z_{th}. Open-circuit the load at terminals a and b as shown. Remove E and replace by a short circuit. The impedance looking into terminals a–b is

$$Z_{th} = Z_2 + \frac{Z_1 Z_3}{Z_1 + Z_3}$$

$$Z_{th} = 1-j2 + \frac{(3+j2)\,(2-j2)}{5}$$

$$Z_{th} = 3-j2.4$$

step 2: Determine E_{oc}. Refer to the original circuit. The load is open-circuited at terminals a–b. Find the voltage across these output terminals.

$$E_{oc} = E\frac{Z_3}{Z_1 + Z_3} = 10\,\frac{2-j2}{5}$$

$$E_{oc} = 4-j4$$

step 3:   The Thevenin's equivalent circuit between terminals *a–b* is

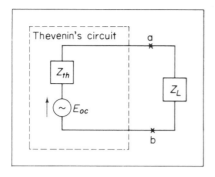

Norton's theorem

Norton's theorem is utilized for current sources applied to shunt elements. Interchanging current and voltage sources permits the application of either Thevenin's or Norton's theorem. Again, the theorem is single-frequency applicable. The theorem is demonstrated in a step-by-step manner. The circuit used is shown in Fig. 16–5.

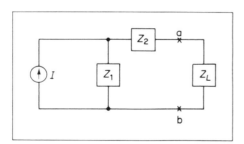

figure 16–5
circuit for Norton's theorem

step 1

Open-circuit the load Z_L at terminals *a–b* and remove the load completely.

step 2

Short-circuit terminals *a–b* and determine the short circuit current I_{sc}.

$$I_{sc} = I \frac{Z_1}{Z_1 + Z_2}$$

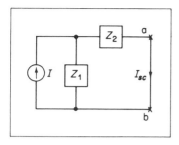

step 3

Remove the short between terminals a–b. Determine Z_{ab}. The following rules are used.

Rule 1: All voltage sources are removed and replaced by their internal impedances or short circuit.

Rule 2: All current sources are open-circuited.

$$Z_{a-b} = Z_2 + Z_1$$

step 4

Construct Norton's equivalent circuit between terminals a–b. Thus,

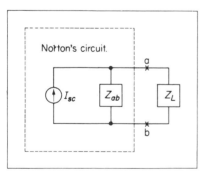

maximum power transfer theorem

The maximum power theorem as applied to ac series circuits states that when the source impedance contains a resistive and reactive component, the load required for maximum power transfer is the complex conjugate of the source impedance. Let the source impedance be equal to $Z_g = R_g + jX_g$. Maximum power will be delivered to the load when the load impedance is the conjugate of the source impedance, or $Z_L = R_g - jX_g$.

An illustrative problem will demonstrate the theory.

sample problem

Given the circuit shown, find:

(a) Z_L required for maximum power transfer.
(b) Power at the load, using Z_L.

$$E = 10\underline{/0^\circ} \qquad X_c = 8\,\Omega$$
$$R = 4\,\Omega \qquad X_L = 4\,\Omega$$

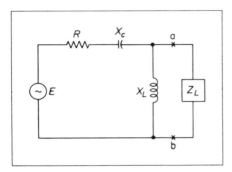

solution

step 1: Thevenize the circuit. Calculate Z_{th}.

$$Z_{th} = \frac{jX_L(R-jX_c)}{R+j(X_L-X_c)}$$

$$Z_{th} = \frac{j4(4-j8)}{4+j(-4)}$$

$$Z_{th} = 2+j3$$

step 2: It is evident that Z_L is the complex conjugate of Z_{th}. Thus,

$$Z_L = 2-j3$$

step 3: Calculate E_{oc}.

$$E_{oc} = E\frac{jX_L}{R+j(X_L-X_c)}$$

$$E_{oc} = 10\frac{j4}{4-j4}$$

$$E_{oc} = -5+j5$$

step 4:   The new circuit is

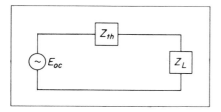

Calculate P_0.

$$P_{\text{load}} = \frac{E_{oc}^2}{4R_L} = \frac{50}{4 \times 2}$$

$$P_{\text{load}} = 6.25 \text{ watts.}$$

T to π and π to T conversions

The π to T and the T to π conversions for ac circuits are identical with the dc circuits, provided that the term *impedance* is substituted for the word *resistance*. The general equations required to convert from π to T as shown in Fig. 16–6 are

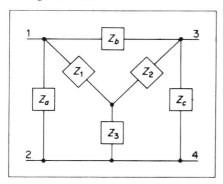

figure 16–6
π to T and T to π conversions

π to T:

$$Z_1 = \frac{Z_a Z_b}{Z_a + Z_b + Z_c}$$

$$Z_2 = \frac{Z_b Z_c}{Z_a + Z_b + Z_c}$$

$$Z_3 = \frac{Z_a Z_c}{Z_a + Z_b + Z_c}$$

T to π:

$$Z_a = \frac{Z_1Z_2+Z_1Z_3+Z_2Z_3}{Z_2}$$

$$Z_b = \frac{Z_1Z_2+Z_1Z_3+Z_2Z_3}{Z_3}$$

$$Z_c = \frac{Z_1Z_2+Z_1Z_3+Z_2Z_3}{Z_1}$$

An illustrative problem will demonstrate the theory.

---

sample problem

The following circuit is given. Find I

where $E = 10\underline{/0°}$ $Z_3 = 3-j4$

       $Z_1 = 4-j3$ $Z_4 = -j4$

       $Z_2 = j2$ $Z_L = 4+j2$

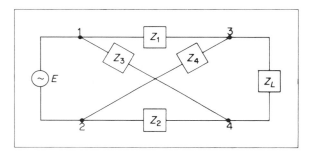

solution

step 1: Redraw the circuit as shown in the figure.

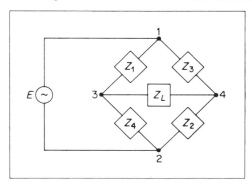

step 2:   Convert π formed by Z_L, Z_4, and Z_2 to an equivalent T.

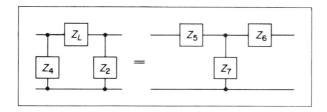

$$Z_5 = \frac{Z_4 Z_L}{Z_4 + Z_L + Z_2}$$

$$Z_5 = \frac{(-j4)(3+j2)}{4}$$

$$Z_5 = 2 - j3$$

$$Z_6 = \frac{Z_L Z_2}{Z_4 + Z_L + Z_2} = \frac{(4+j2)(j2)}{4}$$

$$Z_6 = -1 + j2$$

$$Z_7 = \frac{Z_2 Z_4}{4} = \frac{(j2)(-j4)}{4}$$

$$Z_7 = 2$$

step 3:   The new equivalent circuit is

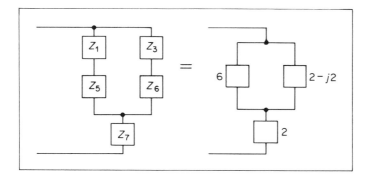

$$Z_T = \frac{6(2-j2)}{8-j2}$$

$$Z_T = 1.76 - j1.05$$

step 4:   Calculate I.

$$I = \frac{E}{Z_T} = \frac{10}{1.76 - j1.05}$$

$$I = 4.88\underline{/30.8°}$$

equivalent T representation of complex network

Any complex network containing linear elements may be exactly represented by an equivalent T network using open- and short-circuit measurements.

For a circuit to be exactly equivalent, the impedances measured at each set of terminals must be equal. Consider the four-terminal complex network shown in Fig. 16–7.

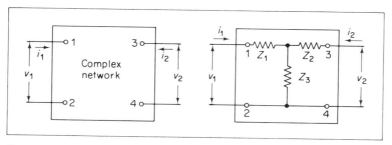

figure 16–7
(a) complex network; (b) equivalent T network

The following nomenclature will be used.

nomenclature

Z_{01} = impedance looking into terminals 1–2 with terminals 3–4 open-circuited.

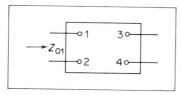

Z_{02} = impedance looking into terminals 3–4 with terminals 1–2 open-circuited.

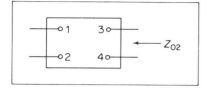

Z_{S1} = impedance looking into terminals 1–2 with terminals 3–4 short-circuited.

Z_{S2} = impedance looking into terminals 3–4 with terminals 1–2 short-circuited.

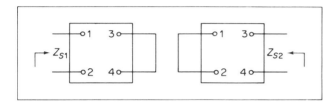

The Z_1, Z_2, and Z_3 arms of a T network that will exactly represent the complex network are determined by these open- and short-circuit measurements. Refer to the T network shown in Fig. 16–8.

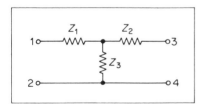

figure 16–8
T network

analysis

$$Z_{01} = Z_1 + Z_3, \qquad Z_{02} = Z_2 + Z_3$$

$$Z_{S1} = Z_1 + \frac{Z_2 Z_3}{Z_2 + Z_3}, \qquad Z_{S2} = Z_2 + \frac{Z_1 Z_3}{Z_1 + Z_3}$$

The cross product of $Z_{01} Z_{S2}$ is equal to

$$Z_{01} Z_{S2} = Z_1 Z_2 + Z_1 Z_3 + Z_2 Z_3$$

The cross product of $Z_{02} Z_{S1}$ is equal to

$$Z_{02}Z_{S1} = Z_1Z_2 + Z_1Z_3 + Z_2Z_3$$

It is evident from these two equations that

$$Z_{01}Z_{S2} = Z_{02}Z_{S1}$$

This means that only three of the four measurements are required to establish circuit requirements and the fourth can be calculated. The product of the two open-circuited impedances results in

$$Z_{01}Z_{02} = Z_1Z_2 + Z_1Z_3 + Z_2Z_3 + Z_3^2$$

Substituting and simplifying yields the following solution for Z_3.

$$Z_3 = \sqrt{Z_{01}Z_{02} - Z_{S1}Z_{02}}$$

Once Z_3 has been determined, the arms Z_1 and Z_2 are found from the following formulas.

$$Z_1 = Z_{01} - Z_3$$
$$Z_2 = Z_{02} - Z_3$$

It should be noted that since Z_3 is a radical quantity, there are two possible solutions to the equivalent network representation. Both solutions will maintain the impedance relationships, but a 180° phase ambiguity will exist. For problem solutions in this text, the positive radical will be assumed valid for all problems.

An illustrative problem will demonstrate the theory.

---

sample problem

A four-terminal network has the following measurements.

$Z_{01} = 60 + j30,$ $Z_{02} = 20 + j40,$ $Z_{S1} = 20 + j30$

Find the equivalent T.

---

solution

step 1:   Calculate Z_3.

$$Z_3 = \sqrt{Z_{02}(Z_{01} - Z_{S1})}$$
$$Z_3 = (20 + j40)\,(40)$$
$$Z_3 = 42.4\underline{/31.75°} = 36 + j22.3$$

step 2: Calculate Z_1.

$$Z_1 = Z_{01} - Z_3$$
$$Z_1 = 60 + j30 - 36 - j22.3$$
$$Z_1 = 24 + j7.7$$

step 3: Calculate Z_2.

$$Z_2 = Z_{02} - Z_3 = 20 + j40 - 36 - j22.3$$
$$Z_2 = -16 + j17.7$$

step 4: The equivalent circuit is

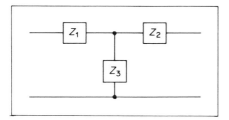

four-terminal network parameters

A four-terminal network may be represented by a "black box," as shown in Fig. 16–9.

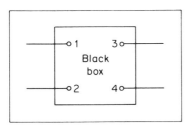

figure 16–9
linear four-terminal network

A mathematical model can be developed based on the combination of voltages and currents. Assume the following set of parameters.

$$\left. \begin{array}{l} v_1 = f(i_1, i_2) \\ v_2 = f(i_1, i_2) \end{array} \right\} \text{ 1 set}$$

Consequently, the required equations are

$$v_1 = i_1 z_{11} + i_2 z_{12}$$
$$v_2 = i_1 z_{21} + i_2 z_{22}$$

The definitions of the "z parameters" are

$$z_{11} = \left.\frac{dv_1}{di_1}\right|_{i_2 = K}$$ input impedance with the output terminals open-circuited

$$z_{12} = \left.\frac{dv_1}{di_2}\right|_{i_1 = K}$$ transfer impedance with the input terminals open-circuited

$$z_{21} = \left.\frac{dv_2}{di_1}\right|_{i_2 = K}$$ transfer impedance with the output terminals open-circuited

$$z_{22} = \left.\frac{dv_2}{di_2}\right|_{i_1 = K}$$ output impedance with the input terminals open-circuited

It is evident from the definitions that each measurement requires open-circuited terminals; therefore, the name "open-circuit parameters."

A single-generator equivalent T network can be derived from the given set of equations, as shown in Fig. 16–10.[1]

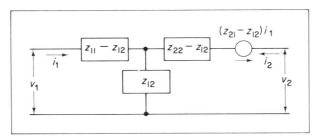

figure 16–10
single-generator equivalent T circuit

A special case exists where $z_{12} = z_{21}$. The voltage generator in the equivalent circuit becomes equal to zero; the new equivalent circuit is shown in Fig. 16–11.

Another mathematical model results when the following set of equations is assumed.

$$\left.\begin{aligned} i_1 &= v_1 y_{11} + v_2 y_{12} \\ i_2 &= v_1 y_{21} + v_2 y_{22} \end{aligned}\right\} \text{1 set}$$

[1]For derivation, see Zeines, *Introduction to Network Analysis* (Englewood Cliffs, N.J.: Prentice-Hall, Inc., 1967), Chapter 1.

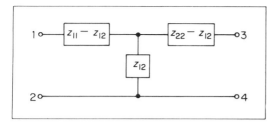

figure 16–11
equivalent T network when $z_{12} = z_{21}$

The definitions of the "y parameters" are

$$y_{11} = \left.\frac{di_1}{dv_1}\right|_{v_2 = K}$$
    input admittance with the output terminals short-circuited

$$y_{12} = \left.\frac{di_1}{dv_2}\right|_{v_1 = K}$$
    transfer admittance with the input terminals short-circuited

$$y_{21} = \left.\frac{di_2}{dv_1}\right|_{v_2 = K}$$
    transfer admittance with the output terminals short-circuited

$$y_{22} = \left.\frac{di_2}{dv_2}\right|_{v_1 = K}$$
    output admittance with the input terminals short-circuited

It is evident from the definitions that each measurement requires short-circuited terminals; therefore, the name "short-circuit measurements."

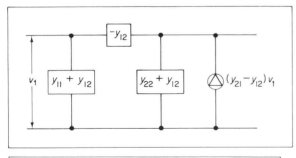

figure 16–12
single-generator π circuit

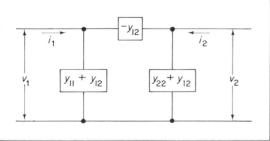

figure 16–13
y parameter circuit

A single-generator π equivalent circuit can be derived from this given set of equations, as shown in Fig. 16–12. A special case exists where $y_{12} = y_{21}$. The current generator becomes open-circuited under these conditions, and the resultant equivalent circuit is shown in Fig. 16–13.

Another mathematical model results with the following given set of equations.

$$v_1 = i_1 h_{11} + v_2 h_{12}$$

$$i_2 = i_1 h_{21} + v_2 h_{22}$$

The definitions of the h parameters are

$h_{11} = \dfrac{dv_1}{di_1}\bigg|_{v_2 = K}$ input impedance with the output terminals short-circuited

$h_{12} = \dfrac{dv_1}{dv_2}\bigg|_{i_1 = K}$ reverse voltage amplification factor with input terminals open-circuited

$h_{21} = \dfrac{di_2}{di_1}\bigg|_{v_2 = K}$ forward current amplification factor with the output terminals short-circuited

$h_{22} = \dfrac{di_2}{dv_2}\bigg|_{i_1 = K}$ output admittance with the input terminals open-circuited

table 16–1   interrelationship of network parameters

↓ To \ From →	z		y		h	
z	z_{11}	z_{12}	$\dfrac{y_{22}}{\Delta y}$	$\dfrac{-y_{12}}{\Delta y}$	$\dfrac{\Delta h}{h_{22}}$	$\dfrac{h_{12}}{h_{22}}$
	z_{21}	z_{22}	$\dfrac{-y_{21}}{\Delta y}$	$\dfrac{y_{11}}{\Delta y}$	$\dfrac{h_{21}}{h_{22}}$	$\dfrac{1}{h_{22}}$
y	$\dfrac{z_{22}}{\Delta z}$	$\dfrac{-z_{12}}{\Delta z}$	y_{11}	y_{12}	$\dfrac{1}{h_{11}}$	$\dfrac{-h_{12}}{h_{11}}$
	$\dfrac{-z_{21}}{\Delta z}$	$\dfrac{z_{11}}{\Delta z}$	y_{21}	y_{22}	$\dfrac{h_{21}}{h_{11}}$	$\dfrac{\Delta h}{h_{11}}$
h	$\dfrac{\Delta z}{z_{22}}$	$\dfrac{z_{12}}{z_{22}}$	$\dfrac{1}{y_{11}}$	$\dfrac{-y_{12}}{y_{11}}$	h_{11}	h_{12}
	$\dfrac{-z_{21}}{z_{22}}$	$\dfrac{1}{z_{22}}$	$\dfrac{y_{21}}{y_{11}}$	$\dfrac{\Delta y}{y_{11}}$	h_{21}	h_{22}

$\Delta z = z_{11}z_{22} - z_{12}z_{21}$
$\Delta y = y_{11}y_{22} - y_{12}y_{21}$
$\Delta h = h_{11}h_{22} - h_{12}h_{21}$

It is evident from the definitions that the measurements are both open- and short-circuit types and are called *hybrid parameters*. A two-generator equivalent circuit can be derived from the given set of equations, as shown in Fig. 16–14.

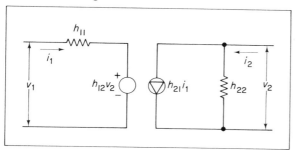

figure 16–14
two-generator equivalent circuit

Since all mathematical models are derived on the basis of currents and voltages existent at the input and output terminals, the sets of parameters are interrelated. Table 16–1 shows the interrelationship of the z, y, and h parameters.

problems

1. Given the circuit shown, determine I_1, using loop circuit analysis.

$E_1 = 10\underline{/0°}$ $Z_2 = 2-j3$
$E_2 = 20\underline{/0°}$ $Z_3 = -j3$
$Z_1 = j5$ $Z_4 = 2-j2$

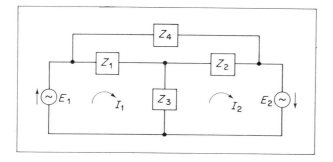

2. Solve for I_1 in the circuit shown at top of page 336.
$E_1 = 10\underline{/0°}$, $Z_1 = 3+j5$, $Z_3 = 4-j4$
$E_2 = 5\underline{/0°}$, $Z_2 = 1+j4$, $Z_4 = 2-j5$

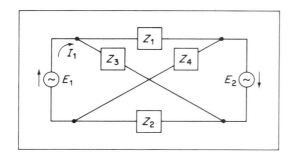

3. Given the circuit shown, determine I_1, using loop circuit analysis.

$E_1 = 10\underline{/0°}, \qquad Z_1 = 3-j4, \qquad Z_4 = j2$

$E_2 = 5\underline{/0°}, \qquad Z_2 = j2, \qquad Z_5 = j3$

$E_3 = 2\underline{/0°}, \qquad Z_3 = 2-j5, \qquad Z_6 = 4-j5$

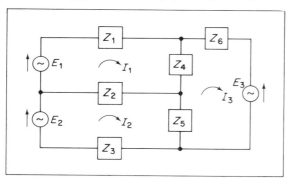

4. Solve for V_a, using nodal analysis.

$E_1 = 10\underline{/0°}, \qquad Z_1 = 5, \qquad Z_4 = 3+j3$

$E_2 = 5\underline{/0°}, \qquad Z_2 = 3+j4, \qquad Z_5 = j2$

$\qquad\qquad\qquad Z_3 = 2-j1$

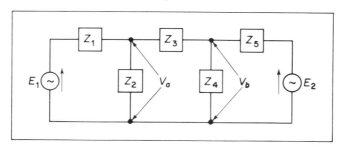

5. Solve for V_a, using nodal analysis.

$I_1 = 1\underline{/0°}, \qquad Y_1 = 2-j3, \qquad Y_3 = j3$

$I_2 = 0.5\underline{/0°}, \qquad Y_2 = 4-j3, \qquad Y_4 = 4+j6$

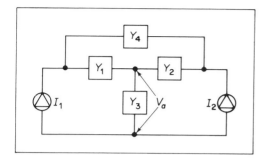

6. Solve for I_L, using Thevenin's theorem.

$$E = 10\underline{/0^\circ} \qquad Z_1 = 2-j4, \qquad Z_4 = 3+j6$$
$$Z_2 = 3+j6, \qquad Z_5 = 3-j2$$
$$Z_3 = 3+j6, \qquad Z_L = 4.25-j1$$

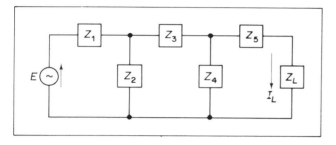

7. Solve for current I_L, using Thevenin's theorem.

$$E = 10\underline{/0^\circ}, \qquad Z_2 = 6+j15, \qquad Z_4 = 6+j15$$
$$Z_1 = 6+j15, \qquad Z_3 = 3-j10, \qquad Z_L = 3-j7.14$$

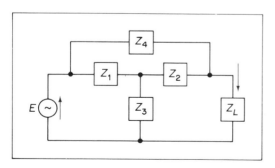

8. Determine the value of Z_L for maximum power transfer and evaluate this power.

$$I_1 = 0.5\underline{/0^\circ}, \qquad I_2 = 0.2\underline{/0^\circ}$$
$$I_3 = 0.3\underline{/0^\circ}, \qquad Z = 4-j6$$

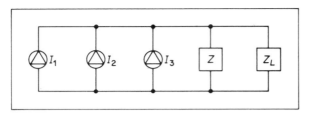

9. Determine the value of Z_L for maximum power transfer and evaluate this power.

$$E = 10\underline{/0°}, \qquad Z_1 = 3-j9, \qquad Z_3 = 2+j6$$
$$Z_2 = 3-j9, \qquad Z_4 = 3-j9$$

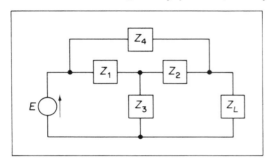

10. Solve for Z_L, using the maximum power transfer theorem.

$$E = 10\underline{/0°}, \qquad Z_1 = 2+j4, \qquad Z_3 = 8+j6$$
$$Z_2 = 8-j6, \qquad Z_4 = 2-j4$$

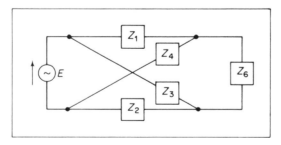

11. Using open- and short-circuit measurements, find the equivalent T for the circuits shown.

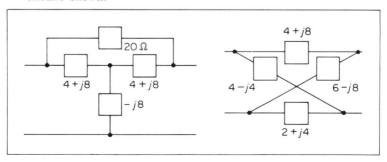

12. A four-terminal network has the following measurements:
$$Z_{01} = 10+j10, \qquad Z_{S1} = 10-j10, \qquad Z_{S2} = 10+j40$$
Find the arms of the equivalent T.

13. A four-terminal network has the following given measurements:
$$Z_{01} = 60+j50, \qquad Z_{02} = 40+j20, \qquad Z_{S1} = 50+j20$$
Find the arms of the equivalent T.

14. A four-terminal network is investigated, and the following measurements are made:
$$z_{11} = 4+j3, \qquad z_{21} = 2+j2.5$$
$$z_{12} = 2+j2.5, \qquad z_{22} = 2.5+j1$$
Find: (a) the y parameters, (b) the h parameters.

15. *Special student problem*
Given the circuit shown, and using open- and short-circuit admittances, determine the elements Y_a, Y_b, and Y_c.

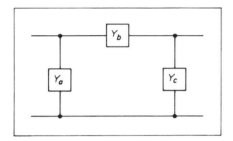

16. In the network shown, determine:
  (a) The value of Z_L required for maximum power transfer.
  (b) The load power, using this value of Z_L.
$$Z_1 = 3+j11, \qquad Z_5 = 9-j5, \qquad Z_9 = 15+j9$$
$$Z_2 = 9-j21, \qquad Z_6 = 4-j8, \qquad Z_{10} = 7-j15$$
$$Z_3 = 9-j21, \qquad Z_7 = 15+j9, \qquad Z_{11} = 1+j1$$
$$Z_4 = 9-j21, \qquad Z_8 = 15+j9, \qquad E = 10\underline{/0^\circ}$$

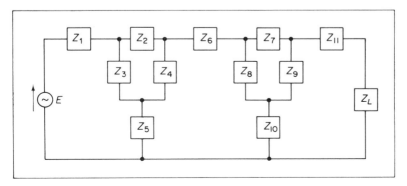

17. Refer to the circuit shown.
$Z_1 = 2-j9,$ $Z_3 = 3+j9,$ $Z_5 = 2-j9$
$Z_2 = 3+j9,$ $Z_4 = 3+j9,$ $Z_6 = 2-j9$
Find: (a) the y parameters, (b) the h parameters.

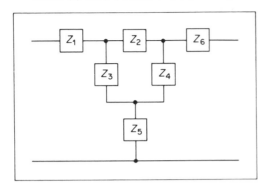

18. Refer to the circuit shown. Find: (a) the y parameters, (b) the h parameters.

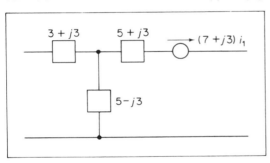

17

transformers

mutual induction

In Chapter 8, it was observed that the voltage across and the current flowing through an inductor were related by the mathematical expression

$$e = L \frac{di}{dt}$$

The term L is called the "self-inductance" of the coil.

When two or more coils are wound and held close together, a mutual inductance is said to exist between the two coils. This mutual inductance is due to a common magnetic flux linked between the two coils. It is evident that the coils must be held so that the lines of flux will link. In addition, power must have been applied and current must be flowing.

Figure 17–1 shows a circuit demonstrating the mutual inductance between the two coils. The generation of an emf in one winding due to the time rate change of current in the other winding is called "mutual induction."

analysis

Assume the secondary terminals to be as shown in Fig. 17–2. For simplicity of analysis, the total flux established in both the primary and secondary each will be separated into two components. The following terminology will be used:

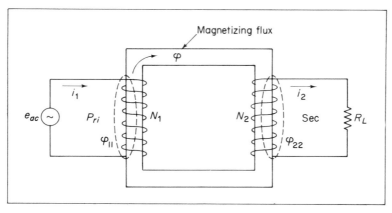

figure 17–1
mutual induction

φ_1 = total flux created by current i_1
φ_2 = total flux created by the current i_2
φ_{11} = leakage flux of the primary. That portion of φ_1 that links only with the turns of the primary circuit
φ_{22} = leakage flux of the secondary. That portion of φ_2 that links only with the turns of the secondary circuit
φ_{12} = magnetizing flux that links the secondary to the primary
φ_{21} = magnetizing flux that links the primary to the secondary.

The total flux in both the primary and secondary circuits is given by

$$\varphi_1 = \varphi_{11} + \varphi_{12} \quad \text{and} \quad \varphi_2 = \varphi_{22} + \varphi_{21}$$

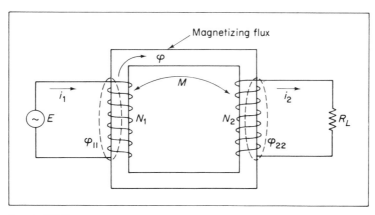

figure 17–2
diagram of a transformer

The magnitude of the emf induced in the primary due to the time rate change of current and flux in the primary is given by two basic laws:

Faraday's Law

$$e_1 = -N_1 \frac{d\varphi_1}{dt}$$

Lenz's Law

$$e_1 = -L_1 \frac{di_1}{dt}$$

where

$\dfrac{d\varphi_1}{dt}$ = change in the lines of force in the primary of the transformer core with respect to time

The magnitude of the emf induced in the secondary is evaluated in a similar manner. Thus,

$$e_2 = -N_2 \frac{d\varphi_2}{dt} \quad \text{and} \quad e_2 = -L_2 \frac{di_2}{dt}$$

It is evident that the self-inductance of a coil can be evaluated in terms of the number of turns and the ratio of the change in flux with respect to the change in current, since

$$e_1 = -L_1 \frac{di_1}{dt} = -N_1 \frac{d\varphi_1}{dt}$$

Solve for L_1. The resultant value is

$$L_1 = N_1 \frac{d\varphi_1}{di_1}$$

Similarly, the value for L_2 is

$$L_2 = N_2 \frac{d\varphi_2}{di_2}$$

The mutual inductance of a circuit can be evaluated by an analogous technique. The symbol representing mutual inductance is the uppercase letter M. Thus,

$$e_1 = -M_{12} \frac{di_2}{dt} \quad \text{and} \quad M_{12} = N_2 \frac{d\varphi_{12}}{di_1} \, H$$

$$e_2 = -M_{21} \frac{di_1}{dt} \quad \text{and} \quad M_{21} = N_1 \frac{d\varphi_{21}}{di_2} \, H$$

The coefficient of coupling between secondary and primary is given by

$$k_{21} = \frac{\varphi_{21}}{\varphi_2}$$

The coefficient of coupling between the primary and secondary is given by

$$k_{12} = \frac{\varphi_{12}}{\varphi_1}$$

The following relationships are substituted for the change in fluxes. Thus,

$$d\varphi_{12} = k_{12} \, d\varphi_1 \quad \text{and} \quad d\varphi_{21} = k_{21} \, d\varphi_2$$

Assume that the mutual inductance between the two coils is equal. That is,

$$M_{12} = M_{21}$$

Then

$$M \, di_1 = k_{12} N_2 \, d\varphi_1$$

and

$$M \, di_2 = k_{21} N_1 \, d\varphi_2$$

Multiplying these two quantities together yields

$$M^2 = \underbrace{k_{12} k_{21}}_{k^2} \underbrace{\left[N_1 \frac{d\varphi_1}{di_1} \right]}_{L_1} \underbrace{\left[N_2 \frac{d\varphi_2}{di_2} \right]}_{L_2}$$

Substituting in the values shown below each quantity results in:

$$M = k \sqrt{L_1 L_2}$$

An illustrative problem will demonstrate the theory.

sample problem

Given the circuit shown, determine the mutual inductance of the transformer,

where: $N_1 = 50$ turns $\qquad L_2 = 800$ mH
$\qquad N_2 = 200$ turns $\qquad k = 0.6$
$\qquad L_1 = 200$ mH

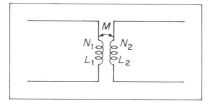

solution

step 1: Calculate M.

$$M = k\sqrt{L_1 L_2}$$
$$M = 0.6\sqrt{200 \times 800 \times 10^{-6}}$$
$$M = 240 \text{ mH}$$

transformation ratio

In an ideal transformer, all of the flux created by the primary current flow links the secondary winding and vice versa. Consequently, the ratio of the induced voltages primary to secondary is

$$\frac{e_1}{e_2} = \frac{N_1}{N_2} = \frac{1}{a}$$

The transformation ratio N_2/N_1 usually is represented by the letter symbol a. The transformation ratio provides a means of transforming the input voltage to any required value with minimum loss. Thus, $e_2 = ae_1$.

In an ideal transformer, the input and output powers are equal. Since $e_1 i_1 = e_2 i_2$, the ratio of the currents can also be defined in terms of the transformation ratio as

$$i_1 = ai_2$$

The driving point impedance of a network is defined as the ratio of the input voltage to the input current. Thus,

$$Z_{dp} = \frac{e_1}{i_1}$$

Connecting a load across the secondary of the transformer causes an additional primary component of current to be produced. The load impedance *reflects* a load into the primary. Note that

$$e_2 = ae_1$$

and

$$i_1 = ai_2$$

The ratio of the output voltage to the output current defines the load impedance. Thus,

$$\frac{e_2}{e_1} = Z_L = \frac{ae_1}{\frac{i_1}{a}} = a^2 Z_{dp}$$

It is evident that transformers can be used to match to any desired impedance. An illustrative problem will demonstrate the theory.

---

sample problem

A transformer is connected to a 4-Ω load. Determine the driving point impedance and the input current.

$N_2 =$ 300 turns $Z_L =$ 4 Ω
$N_1 =$ 1200 turns $i_2 =$ 100 ma

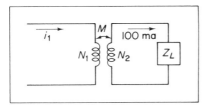

solution

step 1:   Calculate a.

$$a = \frac{N_2}{N_1} = \frac{1}{4}$$

step 2:   Calculate Z_{dp}.

$$Z_{dp} = \frac{Z_L}{a^2} = \frac{4}{1/16} = 4\,\Omega$$

step 3:   calculate i_1.

$$i_1 = ai_2 = \tfrac{100}{4} = 25 \text{ ma}$$

---

sample problem

An audio transformer has 1000 turns in the primary winding. How many turns must be wound in the secondary winding to make a 4-Ω loudspeaker appear as a primary impedance of 3000 ohms?

---

solution

step 1:   Calculate a.

$$Z_L = a^2 Z_{dp}$$

$$\frac{1}{a^2} = \frac{3000}{4} = 750$$

$$\frac{1}{a} = 27.4$$

step 2:   Calculate N_2.

$$N_2 = \frac{N_1}{27.4}$$

$$N_2 = 36.5 \text{ turns}$$

polarity of coils

For the transformer diagram shown in Fig. 17–3, the direction of the coil winding about a core is not indicated. In any given circuit, the voltage developed through the mutual inductance may either aid or oppose the voltage of self-inductance. Since the sign of the mutual inductance can be either positive or negative, a dot notation was developed by the manufacturer to specify the relative polarities of the two coils, as shown in Fig. 17–3.

The following rules will define the polarity of the mutual term as a function of the dots placed on the transformer.

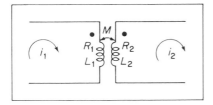

figure 17–3
polarity of the transformer

Rule 1: If the two currents both enter the dot terminals, the polarity of the mutual coupled term will be positive.

Rule 2: If one current leaves the dot terminal while the other current is entering the dot terminal, the polarity of the mutual coupled term is negative.

Once the proper polarity markings and the current directions are established, the correct sign for the induced voltage can be determined. Refer to Fig. 17–4.

Note that the voltage due to the mutual inductance in this case is positive. The dot technique is extremely useful in the case where two coils or more are mutually coupled.

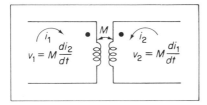

figure 17–4
polarity of mutual inductance

transformer analysis

Consider the transformer circuit shown in Fig. 17–5. The required loop equations are

$$e = R_1 i_1 + L_1 \frac{di_1}{dt} - M \frac{di_2}{dt}$$

$$0 = -M \frac{di_1}{dt} + R_2 i_2 + L_2 \frac{di_2}{dt}$$

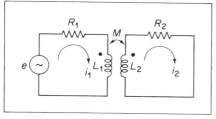

figure 17–5
transformer circuit

Let

$$i_1 = I_{m_1} \sin \omega t \quad \text{and} \quad i_2 = I_{m_2} \sin \omega t$$

Then

$$L_1 \frac{di_1}{dt} = \omega L_1 I_{m_1} \cos \omega t$$

$$M \frac{di_1}{dt} = \omega M I_{m_1} \cos \omega t$$

$$L_2 \frac{di_2}{dt} = \omega L_2 I_{m_2} \cos \omega t$$

$$M \frac{di_2}{dt} = \omega M I_{m_2} \cos \omega t$$

The term ωM is called the "mutual reactance" term and is defined as the ratio of the *voltage* due to mutual inductance to the excitation *current*. The expression for the voltage due to the mutual inductance, therefore, indicates that this voltage leads the excitation current by 90°.

The steady-state ac loop equations are then defined as

$$E = I_1(R_1 + j\omega L_1) - I_2 j\omega M$$

$$0 = -I_1 j\omega M + I_2(R_2 + j\omega L_2)$$

An illustrative problem will demonstrate the theory.

sample problem

The constants for the network shown are

$$R_1 = 400 \ \Omega, \qquad L_1 = 2 \ \text{mH}, \qquad E = 100\underline{/0°}$$
$$R_2 = 600 \ \Omega, \qquad L_2 = 8 \ \text{mH}, \qquad \omega = 10^5 \ \text{rad/sec}$$
$$k = 0.5$$

Find the input impedance of the transformer.

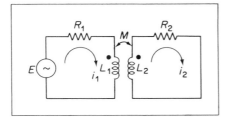

solution

step 1: Calculate M.

$$M = k\sqrt{L_1 L_2}$$
$$M = 0.5 \times 4 \text{ mH} = 2\text{mH}$$

step 2: Write the loop equations.

$$E = I_1(R_1 + j\omega L_1) - I_2(j\omega M)$$
$$0 = -I_1(j\omega M) + I_2(R_2 + j\omega L_2)$$

step 3: Solve for I_1. Use determinants.

$$I_1 = \frac{E(R_2 + j\omega L_2)}{(R_1 + j\omega L_1)(R_2 + j\omega L_2) - (j\omega M)^2}$$

step 4: The input impedance is defined as the ratio of the input voltage to the input current. Thus,

$$Z_{in} = \frac{E}{I_1}$$

$$Z_{in} = \frac{(R_1 + j\omega L_1)(R_2 + j\omega L_2) - (j\omega M)^2}{R_2 + j\omega L_2}$$

Substituting numbers yields the resultant solution. Thus,

$$Z_{in} = \frac{(400 + j200)(600 + j800) + (200)^2}{600 + j800}$$

$$Z_{in} = 400 + j200 + \frac{4 \times 10^4}{200(3 + j4)}$$

$$Z_{in} = 400 + j200 + 24 - j32$$

$$Z_{in} = 424 + j168$$

transformer equivalent circuits

In the analysis of the transformer coupled amplifier, it may be necessary to evaluate the load required for maximum power transfer. The analysis of the transformer coupled audio amplifier is tremendously simplified by using the equivalent T representation of the transformer.

Consider the transformer as a four-terminal network and an equivalent T, as shown in Fig. 17–6.

The arms of the equivalent T can be determined by utilizing the technique of open- and short-circuit measurements discussed in Chapter 16, "Analysis Techniques." The open- and short-circuit measurements on the transformer are

$$Z_{01} = R_1 + jX_{L_1}$$

$$Z_{02} = R_2 + jX_{L_2}$$

$$Z_{S1} = R_1 + jX_{L_1} - \frac{(jX_m)^2}{R_2 + jX_{L_2}}$$

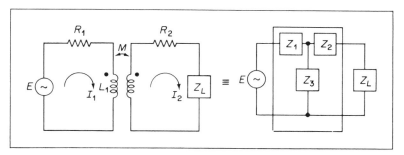

figure 17–6
transformer circuit and its equivalent T network

The actual arms of the equivalent T are

$$Z_3 = \sqrt{Z_{02}(Z_{01} - Z_{S1})}$$

$$Z_3 = \sqrt{(R_2 + jX_{L_2})\left[R_1 + jX_{L_1} - R_1 - jX_{L_1} + \frac{j^2 X_m^2}{R_2 + jX_{L_2}}\right]}$$

$$Z_3 = jX_m$$

The other two arms of the equivalent T can also be determined by

$$Z_1 = Z_{01} - Z_3$$

$$Z_1 = R_1 + j(X_{L_1} - X_m)$$

and

$$Z_2 = Z_{02} - Z_3$$

$$Z_2 = R_2 + j(X_{L_2} - X_m)$$

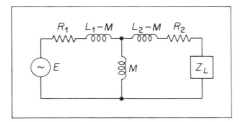

figure 17–7
equivalent T network

The equivalent T network that results from the analytical development is shown in Fig. 17–7.

An illustrative problem will demonstrate the theory.

---

sample problem

The following data are given for a transformer.

$$R_1 = 300 \ \Omega, \qquad L_1 = 3 \ \text{mH}, \qquad \omega = \ 10^5 \ \text{rad/sec}$$
$$R_2 = 150 \ \Omega, \qquad L_2 = 6 \ \text{mH}, \qquad E = 100\underline{/0^\circ}$$

Find: (a) Z_L required for maximum power transfer.
      (b) The output load power.

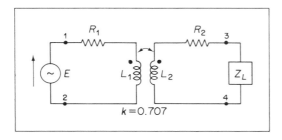

solution

step 1:   Calculate M.

$$M = k \sqrt{L_1 L_2}$$
$$M = 0.707 \ (4.242) = 3 \ \text{mH}$$

step 2:   Convert transformer to an equivalent T, as shown.

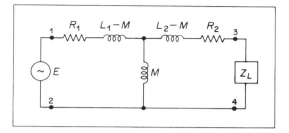

step 3: Calculate Z_{a-b}.

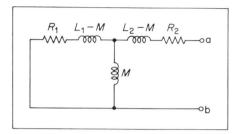

$$Z_{a-b} = R_2 + j\omega(L_2 - M) + \frac{(j\omega M)(R_1 + j(\omega L_1 - \omega M))}{R_1 + j\omega L_1}$$

$$Z_{a-b} = 150 + j300 + \frac{(j300)(150)}{150 + j300}$$

$$Z_{a-b} = 150 + j300 + \frac{j300}{1 + j2}$$

$$Z_{a-b} = 150 + j300 + 120 + j60 = 270 + j360$$

Since Z_L must be equal to the complex conjugate of Z_{a-b}, the value of Z_L is

$$Z_L = 270 - j360$$

step 4: Calculate E_{oc}.

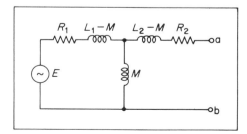

$$E_{oc} = \frac{E(j\omega M)}{R_1 + j\omega L_1} = \frac{100(j300)}{150(1+j2)} = j40(1-j2)$$

$$E_{oc} = 89.6\underline{/26.5°}$$

step 5:   Calculate P_L.

$$P_L = \frac{E_{oc}^2}{4R_L}$$

$$P_L = \frac{(89.6)^2}{4(270)}$$

$$P_L = 7.48 \text{ w}$$

audio transformer

Audio transformers are usually iron core transformers used to match the amplifying device to a speaker. The analysis of the iron core transformer is simplified by converting the transformer to an equivalent T network. The equivalent T network is further simplified by relating all secondary elements to the primary. A circuit requirement is that the driving point impedance remains invariant or constant. It is evident that the secondary elements cannot be placed directly into the primary side, since the input current and the output current are unequal. Consequently, the modification of the secondary elements utilizes the following technique.

1.   The secondary voltage is divided by a.

2.   The secondary current is multiplied by a.

3.   The mutual inductance is divided by a.

4.   All secondary impedances are divided by a^2.

The modified equivalent T circuit configuration of the transformer with all secondary elements related to the primary is shown in Fig. 17–8.

The value usually chosen for a is that value which equates the primary and secondary inductance values. Thus,

$$L_1 - \frac{M}{a} = \frac{L_2}{a^2} - \frac{M}{a}$$

and

$$a^2 = \frac{L_2}{L_1}$$

Since

$$M = k\sqrt{L_1 L_2}$$

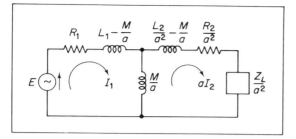

figure 17–8
modified equivalent T circuit of the transformer

then

$$\frac{M}{a} = kL_1$$

Figure 17–8 may be redrawn as shown in Fig. 17–9.

There are many applications in which the iron core transformer can be considered to have definite advantages. In an amplifier circuit, a step-up turns, ratio transformer will increase the output voltage and the resultant stage voltage amplification. An electrical center tap may be inserted on the secondary of the transformer to convert a single-stage amplifier input to a push-pull amplifier stage.

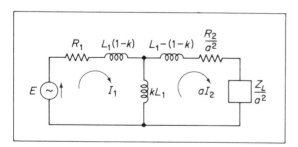

figure 17–9
equivalent circuit in terms of k

transformer testing

A laboratory method that can be used to measure mutual inductance is shown in Fig. 17–10. Two coils, each having a known standard inductance and resistance value, are connected so that the flux produced by the current in the primary is in the same direction as the flux produced in the secondary.

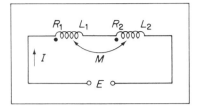

figure 17–10
measurement of mutual inductance (fields aiding)

The coil windings are said to be series-aiding. Using the dot polarity method, we find that the circuit analysis is

$$E = IZ$$

and

$$Z = \sqrt{(R_1 + R_2)^2 + X_T^2}$$

where

$$X_T = X_{L_1} + X_{L_2} + 2X_m$$

Let $L_a = L_1 + L_2 + 2M$ be the total inductance in the circuit with the fields series-aiding. If the connections to the secondary coil are reversed as shown in Fig. 17–11, the fields are series-opposing and the resultant circuit analysis is

$$E = IZ$$

and

$$Z = \sqrt{(R_1 + R_2)^2 + X_T^2}$$

where

$$X_T = X_{L_1} + X_{L_2} - 2X_m$$

Let $L_0 = L_1 + L_2 - 2M$ be the total inductance in the circuit with the fields series-opposing.

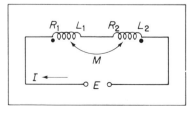

figure 17–11
measurement of mutual inductance (fields opposing)

Subtracting L_0 from L_a yields the following mathematical relationship.

$$L_a = L_1 + L_2 + 2M$$

$$(L_0 = L_1 + L_2 - 2M) \, (-1)$$

Then

$$L_a - L_0 = 4M$$

and

$$M = \frac{L_a - L_0}{4}$$

An illustrative problem will demonstrate the theory.

---

sample problem

Two coils are connected in series across a 200-volt, 400-Hz source. The current that flows is 200 ma. One coil is reversed, and the two coils are connected in series across the source. The current that flows this time is 300 ma.

$R_1 = 200 \, \Omega,$ $R_2 = 300 \, \Omega$

Determine the value of M.

---

solution

step 1:   Calculate Z.

$$Z = \frac{E}{I} = \frac{200}{200} \, 10^3$$

$$Z = 1 \text{k}\Omega$$

step 2:   Calculate L_a.

$$X_{aT} = \sqrt{Z^2 - (R_1 + R_2)^2}$$
$$= \sqrt{(1 - 0.25)10^6}$$
$$= 865 \, \Omega$$

$$L_a \quad = \frac{865}{2\pi \times 400}$$

$$L_a \quad = 345 \text{ mH}$$

step 3:   The seond value of Z is

$$Z = \tfrac{200}{300} \, 10^3 = 667 \, \Omega$$

step 4: Calculate L_0.

$$X_{0T} = \sqrt{Z^2 - (R_1 + R_2)^2}$$

$$X_{0T} = \sqrt{(667)^2 - (500)^2}$$

$$X_{0T} = 444 \ \Omega$$

$$L_0 = \frac{444}{2\pi \times 400}$$

$$L_0 = 176 \ \text{mH}$$

step 5: Calculate M.

$$M = \frac{L_a - L_0}{4}$$

$$M = \frac{(345 - 176)10^{-3}}{4}$$

$$M = 42 \ \text{mH}$$

problems

1. The transformer circuit shown has the following values:

$R_1 = 50 \ \Omega,$ $L_1 = 10 \ \text{mH},$ $k = 0.5$
$R_2 = 100 \ \Omega,$ $L_2 = 40 \ \text{mH},$ $\omega = 10^4 \ \text{rad/sec}$
                                          $E = 100\underline{/0^\circ}$

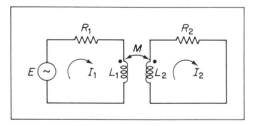

Find I_1.

2. Consider the circuit shown below. The following data are given:

$R_1 = 100 \ \Omega,$ $L_1 = 10 \ \text{mH},$ $\omega = 10^4 \ \text{rad/sec}$
$R_2 = 200 \ \Omega,$ $L_2 = 40 \ \text{mH},$ $E = 100\underline{/0^\circ}$
$R_L = 200 \ \Omega,$ $k = 0.5$

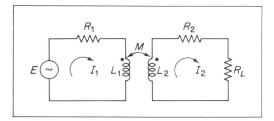

Find: (a) E/I_1, (b) E_0.

3. Find the input current I_1 for the circuit shown.

$R_1 = 100\ \Omega,$ $L_1 = 10$ mH, $k = \frac{2}{3}$

$R_2 = 200\ \Omega,$ $L_2 = 90$ mH, $E_1 = 100\underline{/0°}$

$\omega = 10^4$ rad/sec $E_2 = 50\underline{/0°}$

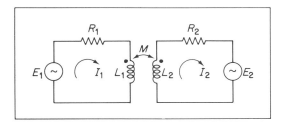

4. The following transformer circuit has the given data.

$R_1 = 10\ \Omega,$ $L_1 = 40$ mH, $\omega = 10^3$ rad/sec

$R_2 = 20\ \Omega,$ $L_2 = 100$ mH, $M = 30$ mH

$R_3 = 30\ \Omega,$ $C = 20\ \mu$F, $E = 100\underline{/0°}$

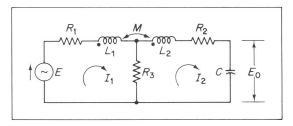

Find: (a) I_1, (b) E_0.

5. The transformer circuit shown has the given data.

$R_1 = 25\ \Omega,$ $L_1 = 25$ mH, $\omega = 10^3$ rad/sec

$R_2 = 50\ \Omega,$ $L_2 = 100$ mH, $k = 0.5$

$R_3 = 50\ \Omega,$ $C = 10\ \mu$F, $E = 100\underline{/0°}$

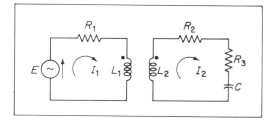

Find: (a) E/I_1, (b) I_2.

6. The transformer circuit shown has the given data.

$R_1 = 20\ \Omega$, $L_1 = 30$ mH, $\omega = 10^3$ rad/sec

$R_2 = 40\ \Omega$, $L_2 = 60$ mH, $M = 20$ mH

$E = 100\underline{/0°}$

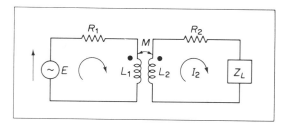

Find: (a) Z_L for maximum output power, (b) P_{out}.

7. The transformer circuit shown has the given data.

$R_1 = 20\ \Omega$, $L_1 = 50$ mH, $\omega = 10^3$ rad/sec

$R_2 = 40\ \Omega$, $L_2 = 60$ mH, $M = 30$ mH

$C = 10\ \mu F$ $E = 100\underline{/0°}$

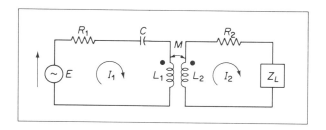

Find: (a) Z_L for maximum output power, (b) P_{out}.

8. The transformer circuit shown has the given data.

$C = 10\ \mu\text{F},$ $R_2 = 100\ \Omega,$ $\omega = 10^3\ \text{rad/sec}$
$R_1 = 100\ \Omega,$ $L_1 = 100\ \text{mH},$ $M = 20\ \text{mH}$
$R = 50\ \Omega,$ $L_2 = 50\ \text{mH}$

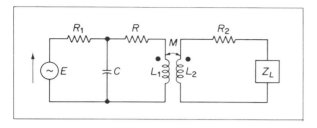

Find: (a) Z_L for maximum P_0, (b) Z_{dp}, using $Z_{L_{\max}}$.

nonsinusoidal waveform analysis

Fourier series

In the preceding sections of the text, the analysis of networks has been principally concerned with the response of a network to a sinusoidal waveform. In this section, the response of a network to a periodic but nonsinusoidal function of time will be analyzed.

A statement that will be made without proof is "Any nonsinusoidal waveform which is periodic, single-valued, and continuous or which has a finite number of discontinuities and has a finite number of maxima and minima in a finite interval of time can be expressed by an infinite series of sine and cosine terms." Thus,

$$f(\alpha) = B_0 + \sum_{k=1,2,3}^{n} \sin k\alpha + \sum_{k=1,2,3}^{n} A_k \cos k\alpha$$

The evaluation of the dc or average component of the nonsinusoidal waveform is determined by multiplying both sides of the equation by $d\alpha$ and integrating over one period of time. Thus,

$$\int_0^{2\pi} f(\alpha)\, d\alpha = \int_0^{2\pi} B_0\, d\alpha + \int_0^{2\pi} B_k \sin k\alpha\, d\alpha + \int_0^{2\pi} A_k \cos k\alpha\, d\alpha$$

The solution to the integration yields the result

$$B_0 = \frac{1}{2\pi} \int_0^{2\pi} f(\alpha)\, d\alpha$$

The evaluation of the Fourier coefficients is determined in a similar manner. Thus, the evaluation of the magnitude of the sine harmonics is performed by multiplying both sides of the original equation by $\sin n\alpha \, d\alpha$ and then integrating over one period of time. Thus,

$$\int_0^{2\pi} f(\alpha) \sin n\alpha \, d\alpha = \int_0^{2\pi} B_0 \sin n\alpha \, d\alpha + \int_0^{2\pi} B_k \sin k\alpha \sin n\alpha \, d\alpha$$

$$+ \int_0^{2\pi} A_k \cos k\alpha \sin n\alpha \, d\alpha$$

$$\int_0^{2\pi} B_0 \sin n\alpha \, d\alpha = \frac{B_0}{n} (-\cos n\alpha) \bigg|_0^{2\pi} = 0$$

$$\int_0^{2\pi} A_k \cos k\alpha \sin n\alpha \, d\alpha = -\frac{A_k}{2} \left[\frac{\cos (n-k)\alpha}{n-k} + \frac{\cos (n+k)\alpha}{n+k} \right]_0^{2\pi}$$

$$A_k = 0 \text{ for all values of } n \text{ and } k$$

Letting $n = k$ yields the integral

$$\int_0^{2\pi} A_k \cos k\alpha \sin k\alpha \, d\alpha = -\frac{A_k}{4k} (\cos 2k\alpha) \bigg|_0^{2\pi} = 0$$

$$\int_0^{2\pi} B_k \sin k\alpha \sin n\alpha \, d\alpha = \frac{B_k}{2} \left[\frac{\sin (n-k)\alpha}{n-k} - \frac{\sin (n+k)\alpha}{n+k} \right]_0^{2\pi} = 0$$

Letting $k = n$ yields the integral

$$\int_0^{2\pi} B_k \sin^2 k\alpha \, d\alpha = \frac{B_k}{2k} \left[k\alpha - \frac{1}{2} \sin 2k\alpha \right]_0^{2\pi} = B_k \pi$$

The value of B_k then becomes equal to

$$B_k = \frac{1}{\pi} \int_0^{2\pi} f(\alpha) \sin k\alpha \, d\alpha$$

The evaluation of the cosine harmonics proceeds in a similar manner to the foregoing discussion. Thus,

$$A_k = \frac{1}{\pi} \int_0^{2\pi} f(\alpha) \cos k\alpha \, d\alpha$$

The required integrals that are to be used are

$$\int \sin k\alpha \sin n\alpha \, d\alpha = \left(\frac{\sin (n-k)\alpha}{2(n-k)} - \frac{\sin (n+k)\alpha}{2(n+k)} \right) \quad k^2 \neq n^2$$

$$\int \sin n\alpha \cos k\alpha \, d\alpha = -\left(\frac{\cos (n-k)\alpha}{2(n-k)} + \frac{\cos (n+k)\alpha}{2(n+k)}\right) \quad k^2 \neq n^2$$

$$\int \cos k\alpha \cos n\alpha \, d\alpha = \left(\frac{\sin (n-k)\alpha}{2(n-k)} + \frac{\sin (n+k)\alpha}{2(n+k)}\right) \quad k^2 \neq n^2$$

$$\int \sin^2 k\alpha \, d\alpha = \frac{1}{2k}(k\alpha - \sin k\alpha \cos k\alpha)$$

$$\int \cos^2 k\alpha \, d\alpha = \frac{1}{2k}(k\alpha + \sin k\alpha \cos k\alpha)$$

$$\int \alpha \cos k\alpha \, d\alpha = \left(\frac{\cos k\alpha}{k^2} + \frac{\alpha \sin k\alpha}{k}\right)$$

$$\int \alpha \sin k\alpha \, d\alpha = \left(\frac{\sin k\alpha}{k^2} - \frac{\alpha \cos k\alpha}{k}\right)$$

The magnitude of the cosine harmonics is evaluated by multiplying both sides of $f(\alpha)$ by $\cos n\alpha \, d\alpha$ and integrating over one period of time. Thus,

$$\int_0^{2\pi} f(\alpha) \cos n\alpha \, d\alpha = \int_0^{2\pi} B_0 \cos n\alpha \, d\alpha + \int_0^{2\pi} B_k \sin k\alpha \cos n\alpha \, d\alpha$$
$$+ \int_0^{2\pi} A_k \cos k\alpha \cos n\alpha \, d\alpha$$

Substituting in the proper definite integrals and evaluating A_k, we find that $A_k = 0$. On the other hand, when $k^2 = n^2$, then the solution for A_k is

$$A_k = \frac{1}{\pi} \int_0^{2\pi} f(\alpha) \cos k\alpha \, d\alpha$$

An illustrative problem will demonstrate the theory.

sample problem

Find the Fourier series for the waveform shown.

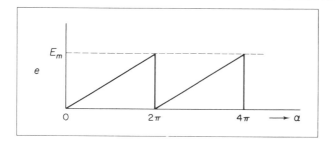

solution

step 1: Determine the equation of the waveform. Thus,

$$f(\alpha) = \frac{E_m}{2\pi} \alpha \Big|_0^{2\pi}$$

step 2: Determine B_0.

$$B_0 = \frac{1}{2\pi} \int_0^{2\pi} f(\alpha)\, d\alpha$$

$$B_0 = \frac{1}{2\pi} \int_0^{2\pi} \frac{E_m}{2\pi} \alpha\, d\alpha$$

$$B_0 = \frac{E_m}{4\pi^2} \left(\frac{\alpha^2}{2}\right) \Big|_0^{2\pi}$$

$$B_0 = \frac{E_m}{2}$$

step 3: Determine the A_k terms.

$$A_k = \frac{1}{\pi} \int_0^{2\pi} f(\alpha) \cos k\alpha\, d\alpha$$

$$A_k = \frac{1 \times E_m}{2\pi^2} \int_0^{2\pi} \alpha \cos k\alpha\, d\alpha$$

The definite integral required is

$$\int \alpha \cos k\alpha\, d\alpha = \left[\frac{\cos k\alpha}{k^2} + \frac{\alpha \sin k\alpha}{k}\right]$$

Thus,

$$A_k = \frac{E_m}{2\pi^2} \left[\frac{\cos k\alpha}{k^2} + \frac{\alpha \sin k\alpha}{k}\right]_0^{2\pi}$$

$A_k = 0$ for all values of k

step 4: Determine the B_k terms.

$$B_k = \frac{1}{\pi} \int_0^{2\pi} f(\alpha) \sin k\alpha\, d\alpha$$

$$B_k = \frac{E_m}{2\pi^2} \int_0^{2\pi} \alpha \sin k\alpha\, d\alpha$$

The definite integral required is

$$\int \alpha \sin k\alpha\, d\alpha = \left(\frac{\sin k\alpha}{k^2} - \frac{\alpha \cos k\alpha}{k}\right)$$

$$B_k = \frac{E_m}{2\pi^2} \left[\frac{\sin k\alpha}{k^2} - \frac{\alpha \cos k\alpha}{k} \right]_0^{2\pi}$$

$$B_k = \frac{-E_m}{\pi k}$$

The Fourier series for the waveform given is

$$f(\alpha) = \frac{E_m}{2} = \frac{E_m}{\pi} \sin \alpha - \frac{E_m}{2\pi} \sin 2\alpha$$

$$- \frac{E_m}{3\pi} \sin 3\alpha - \frac{E_m}{4\pi} \sin 4\alpha - \cdots$$

symmetry

Recognition of whether a function is odd or even may be exceedingly desirable, since it may reduce the amount of labor required to evaluate the general Fourier coefficients. Before we proceed into types of symmetries, it is necessary to establish the axis definitions.

Consider the axis to have four quadrants. The first quadrant, rotating in a counterclockwise manner, defines $f(\alpha)$. The second quadrant defines $f(-\alpha)$. The third and fourth quadrants denote $-f(-\alpha)$ and $-f(\alpha)$, respectively. See Fig. 18–1.

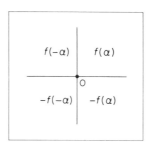

figure 18–1
axis definitions used for symmetries

All symmetries will be established at a reference axis existing at 0, π, 2π, etc. It is, therefore, evident that the value of the ordinate in both the first and second quadrants is positive, whereas it is negative in both the third and fourth quadrants.

It is evident that some waveforms have no B_k terms or A_k terms. Consequently, it may be stated that some functions possess certain symmetries, which indicates that some harmonics do not exist. Recognition of these symmetries will reduce the amount of labor required to evaluate the Fourier coefficients.

A waveform is said to have half-wave symmetry, sometimes called mirror symmetry, when the following condition is met. Thus,

$$f(\alpha) = -f(\alpha + \pi)$$

When a waveform has this type of symmetry, the B_0 term is zero, and there are no even harmonics existent in the waveform. An example of this type of symmetry is given in Fig. 18–2.

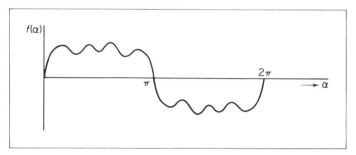

figure 18–2
mirror symmetry

A further test of mirror symmetry is placing the second half of the waveform under the first half and verifying the mirror effect. This effect is shown in Fig. 18–3.

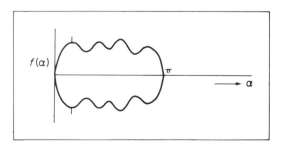

figure 18–3

A waveform is said to have point symmetry when the following condition is satisfied.

$$f(\alpha) = -f(-\alpha)$$

When a waveform has this type of symmetry, the B_0 term is zero, and there are no cosine harmonics in the waveform. This is written mathematically as

$$B_0 = 0 \quad \text{and} \quad A_k = 0 \text{ for all values of } k$$

An example of this type of symmetry is shown in Fig. 18–4.

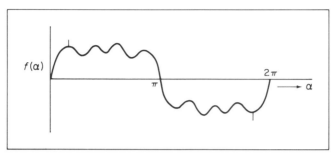

figure 18–4
point symmetry

A waveform is said to have zero axis symmetry when the following test condition is satisfied. Thus,

$$f(\alpha) = f(-\alpha)$$

Under this condition, all of the B_k terms are equal to zero. Thus,

$$B_k = 0$$

An example of this type of symmetry is shown in Fig. 18–5.

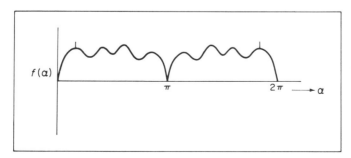

figure 18–5
zero-axis symmetry

A waveform is said to have coincident symmetry when the following test condition is satisfied. Thus,

$$f(\alpha) = f(\alpha + \pi)$$

When this condition exists, the function has no odd harmonics in either the A_k or B_k terms. An example of this type of symmetry is shown in Fig. 18–6.

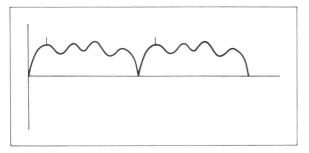

figure 18–6
coincident symmetry

An illustrative example will demonstrate the theory.

sample problem

The following waveform is given. Determine the Fourier series.

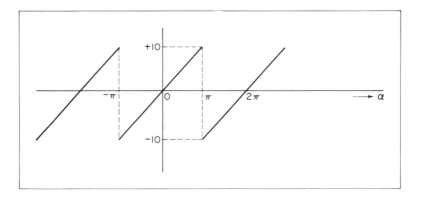

solution

step 1: From the symmetry, it is evident that the $B_0 = 0$. Also, this wave-form satisfies the test that

$$f(\alpha) = -f(-\alpha), \qquad \Sigma A_k = 0$$

step 2: The only terms that exist are the sinusoidal harmonics. Thus,

$$B_k = \frac{1}{\pi} \int_{-\pi}^{2\pi} f(\alpha) \sin k\alpha \, d\alpha$$

The function $f(\alpha)$ is

$$f(\alpha) = \frac{10}{\pi} \alpha \Big|_{-\pi}^{+\pi}$$

step 3:

$$B_k = \frac{10}{\pi^2} \int_{-\pi}^{+\pi} \alpha \sin k\alpha \, d\alpha$$

Using the definite integral given on page 366, the resultant solution is

step 4: $$B_k = \frac{10}{\pi^2} \left[\frac{\sin k\alpha}{k^2} - \frac{\alpha \cos k\alpha}{k} \right]_{-\pi}^{+\pi}$$

$$B_k = -\frac{20}{k\pi} \cos k\pi$$

Note:

$$B_1 = +\frac{20}{\pi}, \qquad B_2 = -\frac{10}{\pi}, \qquad B_3 = \frac{20}{3\pi}, \qquad B_4 = \frac{5}{\pi}$$

step 5: The Fourier series for the waveform is

$$f(\alpha) = \frac{20}{\pi} \sin \alpha - \frac{10}{\pi} \sin 2\alpha + \frac{20}{3\pi} \sin 3\alpha + \cdots$$

or

$$f(\alpha) = \sum_{k=1,2,3}^{n} B_k \sin k\alpha$$

where

$$B_k = -\frac{20}{\pi k} \cos k\pi$$

sample problem

Consider the waveform shown and determine the Fourier series.

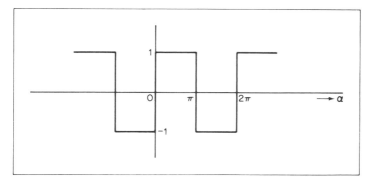

solution

This function has the equation

$$f(\alpha) = 1 \Big|_0^\pi + -1 \Big|_\pi^{2\pi}$$

This waveform has both mirror and point symmetry. Consequently, the following terms can be eliminated.

$$B_0 = 0, \qquad A_k = 0$$

All even harmonics are zero. The odd sine harmonics are the only components existent in this waveform. Thus,

$$B_k = \frac{1}{\pi} \int_0^\pi \sin k\alpha \, d\alpha + \frac{1}{\pi} \int_\pi^{2\pi} -\sin k\alpha \, d\alpha$$

$$B_k = \frac{-1}{\pi} \left[\frac{\cos k\alpha}{k} \right]_0^\pi - \frac{1(-1)}{\pi} \left[\frac{\cos k\alpha}{k} \right]_\pi^{2\pi}$$

$$B_k = \frac{1}{k\pi} (1 - \cos k\pi) - \frac{1}{k\pi} (\cos k\pi - 1)$$

$$B_k = \frac{2}{k\pi} (1 - \cos k\pi)$$

$$B_k = 0 \text{ when } k \text{ is even}, \qquad B_k = \frac{4}{k\pi} \text{ when } k \text{ is odd}$$

The resultant solution for the $f(\alpha)$ is

$$f(\alpha) = \frac{4}{\pi} \left[\sin \alpha + \frac{1}{3} \sin 3\alpha + \frac{1}{5} \sin 5\alpha + \cdots \right]$$

sample problem

The Fourier series of the given waveform is required. Determine the harmonic components.

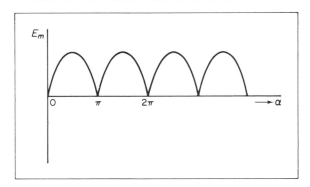

solution

The equation of the waveform is

$$f(\alpha) = E_m \sin \alpha \Big|_0^{\pi} + -E_m \sin \alpha \Big|_{\pi}^{2\pi}$$

This waveform has both zero-axis and coincident symmetry. This indicates that the following terms can be eliminated. Thus,

$$B_k = 0, \qquad \text{all odd harmonics} = 0$$

The only components present in this waveform are the B_0 term and even cosine harmonics. Thus,

$$B_0 = \frac{1}{2\pi} \int_0^{\pi} E_m \sin \alpha \, d\alpha + \frac{-1}{\pi} \int_{\pi}^{2\pi} E_m \sin \alpha \, d\alpha$$

$$B_0 = \frac{-2E_m}{2\pi} (\cos \alpha) \Big|_0^{\pi}$$

$$B_0 = \frac{2E_m}{\pi}$$

The cosine harmonics are determined by

$$A_k = \frac{1}{\pi} \int_0^{\pi} E_m \sin \alpha \cos k\alpha \, d\alpha + \frac{-E_m}{\pi} \int_{\pi}^{2\pi} \sin \alpha \cos k\alpha \, d\alpha$$

$$A_k = -\frac{E_m}{2\pi} \left[\frac{\cos (1-k)\alpha}{(1-k)} + \frac{\cos (1+k)\alpha}{1+k} \right]_0^{\pi}$$

$$+ \frac{E_m}{2\pi} \left[\frac{\cos (1-k)\alpha}{1-k} + \frac{\cos (1+k)\alpha}{1+k} \right]_{\pi}^{2\pi}$$

$$A_k = -\frac{E_m}{2\pi} \left(\frac{\cos (1-k)\pi}{1-k} - \frac{1}{1-k} + \frac{\cos (1+k)\pi}{1+k} - \frac{1}{1+k} \right)$$

$$+ \frac{E_m}{2\pi} \left(\frac{\cos (1-k)2\pi}{1-k} - \frac{\cos (1-k)\pi}{1-k} \right.$$

$$+ \left. \frac{\cos (1+k)2\pi}{1+k} - \frac{\cos (1+k)\pi}{1+k} \right)$$

The sum or difference of two odd numbers results in an even number. This causes A_k to be equal to zero. Thus,

$$A_k = 0 \quad \text{when} \quad k = \text{odd number}$$

The value of A_k when k is equal to an even number is

$$A_k = \frac{4E_m}{\pi(1-k^2)}$$

The equation for the waveform is

$$f(\alpha) = \frac{2E_m}{\pi} + \sum_{k=\text{even}}^{n} \frac{4E_m}{\pi(1-k^2)} \cos k\alpha$$

graphical techniques

It is evident that the previous method of analysis requires that the equation of the function be known before one proceeds on the evaluation of the Fourier coefficients. This technique has been thoroughly demonstrated. However, many functions cannot be expressed explicitly as a function of time. Consequently, a graphical approach is then required.

Consider the waveform shown in Fig. 18–7.

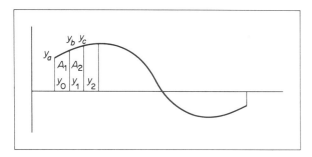

figure 18–7
graphical method of determining Fourier coefficients

For most practical applications, the 12-point schedule method is used. This means that each increment is 30° or $\pi/6$ radians. The trapezoidal rule is the mathematical approach used in this type of waveform. One complete cycle is divided into 12 separate divisions, each having an increment of 30°. The ordinates are numbered from 0 to 11. The ordinates above the axis are positive, and the ordinates below the axis are negative.

The area of the first trapezoid is

$$S_1 = \frac{y_0 \pi}{6}$$

where

$$y_0 = \frac{y_a + y_b}{2}$$

The area of the second trapezoid is

$$S_2 = \frac{y_1 \pi}{6}$$

where

$$y_1 = \frac{y_b + y_c}{2}$$

This procedure is continued until S_{12} has been evaluated. The total area is

$$S_T = S_1 + S_2 + S_3 + \cdots + S_{12}$$

The average or dc component of the waveform is defined as the ratio of the area under the curve with respect to 2π. Thus,

$$B_0 = \frac{S_T}{2\pi}$$

or

$$B_0 = \frac{1}{12} [y_0 + y_1 + y_2 + \cdots + y_{11}]$$

Although the increments may be decreased, the technique does not vary. Thus, a 36-point schedule method can be used where each increment is 10°. The ordinates range from y_0 to y_{35}. The constant in front of the summation of the y ordinates will vary accordingly.

The evaluation of the magnitude of the sine harmonics is performed in a similar manner. Thus, using the 12-point method, we have

$$B_k = \frac{1}{\pi} \int_0^{2\pi} f(\alpha) \sin k\alpha \, d\alpha$$

or

$$B_k = \frac{\text{area under the } f(\alpha) \sin k\alpha \text{ curve}}{\pi}$$

Thus,

$$B_k = \frac{\frac{\pi}{6}(y_0 \sin k0° + y_1 \sin k30° + y_2 \sin k60° + \cdots + y_{11} \sin k330°)}{\pi}$$

From trigonometry, it is evident that

$$\sin k30° = -\sin k330°$$
$$\sin k60° = -\sin k300°$$

and so on.
Then

$$B_k = \tfrac{1}{6} \, [y_0 \sin k0° + (y_1 - y_{11}) \sin k30°$$
$$+ (y_2 - y_{10}) \sin k60° + (y_3 - y_9) \sin k90°$$
$$+ \cdots + y_6 \sin k180°]$$

Using the same technique, we find that the value of A_k is

$$A_k = \tfrac{1}{6} \, [y_0 \cos k0° + y_1 \cos k30° + y_2 \cos k60° + \cdots + y_{11} \cos k330°]$$

From trigonometry, it is evident that

$$\cos k30° = \cos k330°$$
$$\cos k60° = \cos k300°$$

etc.
Then

$$A_k = \tfrac{1}{6} \, [y_0 \cos k0° + (y_1 + y_{11}) \cos k30° + (y_2 + y_{10}) \cos k60°$$
$$+ (y_3 + y_9) \cos k90° + \cdots + y_6 \cos k180°]$$

An illustrative problem will demonstrate the theory.

---

sample problem

The following values are measured in the laboratory. Determine the Fourier series.

α	0	30°	60°	90°	120°	150°	180°
$f(\alpha)$	0	20	40	65	68	50	0
α	210°	240°	270°	300°	330°	360°	
$f(\alpha)$	-50	-68	-65	-40	-20	0	

solution

step 1: Determine the value of the ordinates. Thus,

$$y_0 = 10 \qquad\qquad y_6 = -25$$
$$y_1 = 30 \qquad\qquad y_7 = -59$$
$$y_2 = 52.5 \qquad\quad y_8 = -66.5$$
$$y_3 = 66.5 \qquad\quad y_9 = -52.5$$
$$y_4 = 59 \qquad\qquad y_{10} = -30$$
$$y_5 = 25 \qquad\qquad y_{11} = -10$$

step 2: Calculate the average value of the waveform.

$$B_0 = \tfrac{1}{12}(y_0 + y_1 + y_2 + \cdots + y_{11})$$
$$B_0 = \tfrac{1}{12}(+243 - 243) = 0$$

step 3: Calculate the value of B_k.

$$B_k = \tfrac{1}{6}\{0 + [30 - (-10)]\sin k30°$$
$$+ [52.5 - (-30)]\sin k60°$$
$$+ [66.5 - (-52.5)]\sin k90°$$
$$+ \cdots + (-25)\sin k180°\}$$
$$B_k = \tfrac{1}{6}[40\sin k30° + 82.5\sin k60°$$
$$+ 119\sin k90° + 125.5\sin k120° + 84\sin k150°]$$

step 4: Determine the value of A_k.

$$A_k = \tfrac{1}{6}[y_0\cos k0° + (y_1 + y_{11})\cos k30°$$
$$+ (y_2 + y_{10})\cos k60° + (y_3 + y_9)\cos k90°$$
$$+ \cdots + (y_6)\cos k180°]$$
$$A_k = \tfrac{1}{6}[10 + 20\cos k30° + 22.5\cos k60°$$
$$+ 14\cos k90° + (-7.5)\cos k120°$$
$$+ (-34)\cos k150° - 25\cos k180°]$$

step 5: The individual harmonics can be calculated by inserting the number of the desired harmonic for k and solving. The solution of the various harmonics is left to the student as an exercise.

power considerations

The average power is determined when the waveform is nonsinusoidal by evaluating the power due to each frequency component and then by adding all such values of power. Consequently, when the waveform has both a B_k and an A_k term at the same frequency, these two signal generators must be converted to a single source. Thus, let:

$$B_k = C_k \cos \theta_k$$

$$A_k = C_k \sin \theta_k$$

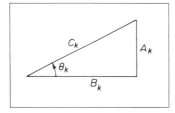

Thus,

$$f(\alpha) = B_0 + \Sigma \, B_k \sin k\alpha + \Sigma \, A_k \cos k\alpha$$

and

$$f(\alpha) = B_0 + C_k \sin k\alpha \cos \theta_k + C_k \cos k\alpha \sin \theta_k$$

$$f(\alpha) = B_0 + \Sigma \, C_k \sin (k\alpha + \theta_k)$$

Note that

$$C_k = \sqrt{A_k^2 + B_k^2}$$

$$\theta_k = \tan^{-1} \frac{A_k}{B_k}$$

The effective or rms value of the waveform is

$$C_{T\text{rms}} = \sqrt{B_0^2 + \frac{C_1^2}{2} + \frac{C_2^2}{2} + \frac{C_3^2}{2} + \cdots}$$

Harmonic distortion analysis utilizes the values of the C terms. Thus, the second harmonic distortion is

$$D_2 = \frac{C_2}{C_1} \, 100\%$$

The percentage of third harmonic distortion is

$$D_3 = \frac{C_3}{C_1} 100\%$$

The total or overall distortion in the system is

$$D_T = \sqrt{D_2^2 + D_3^2 + D_4^2 + D_5^2 + \cdots}$$

problems

1. Develop the Fourier series for the indicated wave.

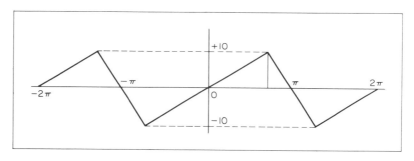

2. Develop the Fourier series for the indicated wave.

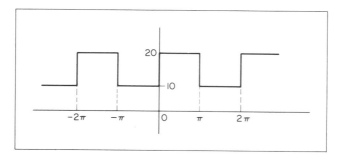

3. Find the Fourier series for the indicated wave.

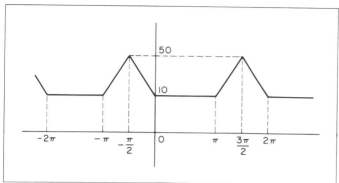

4. Develop the Fourier series for the indicated waveform.

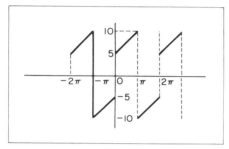

5. Determine the Fourier series for the indicated waveform.

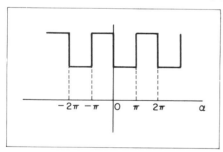

6. Determine the percentage of second and third harmonic distortion.

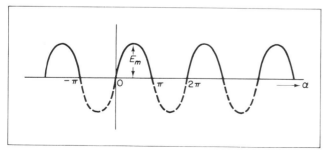

7. Given the waveform shown, determine the average value and the first three harmonics.

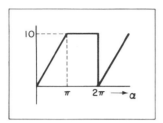

8. Given the waveform shown, determine the complete Fourier series.

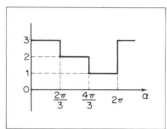

9. Given the waveform shown, determine the average value and the first three harmonics.

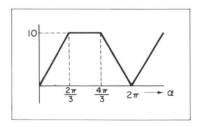

10. Determine the first three harmonics for the waveform shown.

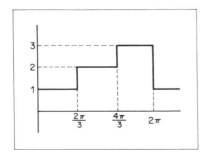

11. Determine the average value and the first three harmonics for the waveform shown.

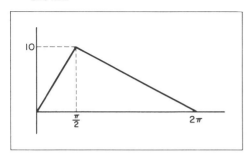

12. A current waveform is defined over one complete cycle by the following data.

α	i	α	i
0	− 2.00	210	− 5.00
30	+ 3.00	240	− 7.66
60	+ 9.66	270	− 9.00
90	+ 13.00	300	− 7.66
120	+ 9.66	330	− 5.00
150	+ 3.00	360	− 2.00
180	− 2.00		

Find: B_0, B_1, B_2, B_3, A_1, A_2, and A_3.

appendices

A

slide rule instructions

introduction

The purpose of the slide rule is to provide a numerical solution of reasonable accuracy with extreme rapidity. For electrical engineering work, the log-log vector slide rule is extremely useful. The minimum number of scales required for electrical technology operations is the following: A, B, C, D, C_1, K, L, S, T, ST.

multiplication (use of C and D scales)

The C and D scales are identical, starting at 1 and ending at 1. The 1 at each end of the scale is referred to as the index. The problem that generally exists lies in properly placing and reading the numbers on these scales. For example, to set number 455 on the D scale, the scale calibration assumes the index at the left to be equal to 100 and the index at the other end to be 1000. Thus, the D scale calibration is from 100 to 1000, and 455 can then be set accordingly. The procedure for multiplying two numbers is

step 1

Place either index of the C scale over one number on the D scale.

step 2

Move the hairline of the glass cursor until it is centered over the second number on the C scale.

step 3

Read the answer under hairline on the *D* scale.

step 4

The decimal point is left up to the student. Using powers of ten on the original problem will help in evaluating the decimal point.

Some illustrative problems will demonstrate the theory.

---

sample problem

Multiply 8 by 2, using the slide rule.

---

solution

step 1: Place the right index of *C* over the number 8 on the *D* scale.

step 2: Move the hairline of the glass cursor to 2 on the *C* scale.

step 3: Read the value of the solution on the *D* scale. In this case, the number is one six, or 16.

---

sample problem

Multiply 7 by 22.

---

solution

step 1: Place the right index of *C* over the number 7 on the *D* scale.

step 2: Move the hairline of the cursor to number 22 on the *C* scale.

step 3: Read the value of the solution on the *D* scale. In this case, the number is 154.

As an exercise, verify the answers to the following:

1. $4 \times 6 = 24$

2. $15 \times 41 = 615$

3. $320 \times 1.6 = 512$

4. $75.3 \times 913 = 68.7 \times 10^4$

5. $12.3 \times 74.5 = 919$

6. $29.2 \times 8.81 = 257$

7. $510 \times 196,247 = 10^7$

8. $3.92 \times 0.0314 = 0.123$

9. $0.00613 \times 0.000746 = 4.57 \times 10^{-8}$

10. $51,200 \times 0.00182 = 60.5$

division

The procedure for dividing two numbers is as follows.

step 1

Set the hairline of the cursor over the numerator on the *D* scale.

step 2

Move the *C* scale so that the denominator lines up with the hairline.

step 3

Move the hairline to either index of the *C* scale and read the answer on the *D* scale.

Some illustrative problems will demonstrate the theory.

---

sample problem

Using the slide rule, divide 380 by 19.

---

solution

step 1: Set the hairline of the cursor over 380 on the *D* scale.

step 2: Move the *C* scale so that the number 19 lines up with the hairline.

step 3: Move the hairline to either index of the *C* scale and read the answer 2 on the *D* scale. Using the powers of ten, we find that the answer is 20.

---

sample problem

Using the slide rule, divide 420 by 1050.

---

solution

step 1: Set the hairline of the cursor over the number 420 on the *D* scale.

step 2: Move the *C* scale so that the number 1050 lines up with the hairline.

step 3: Move the hairline to either index of the *C* scale and read the number 4 on the *D* scale. Using powers of ten, we find that the answer is 0.4.

Verify the solutions to the following problems.

1. $\dfrac{25}{5} = 5$

6. $\dfrac{20,200}{1.01} = 2 \times 10^4$

2. $\dfrac{450}{15} = 30$

7. $\dfrac{0.00925}{283} = 3.27 \times 10^{-5}$

3. $\dfrac{4030}{11.6} = 347$

8. $\dfrac{41,600}{0.0078} = 5.33 \times 10^6$

4. $\dfrac{12}{816} = 0.0146$

9. $\dfrac{63,100}{19.75} = 320$

5. $\dfrac{81.5 \times 10^{-3}}{1.37} = 59.4 \times 10^{-3}$

10. $\dfrac{1240}{0.099} = 1.25 \times 10^5$

reciprocal scale $(C_1$ or $D_1)$

Examination of the C_1 scale reveals that each number on the C_1 scale is the reciprocal of the numbers on the C scale. Thus, the reciprocal of 5 is 0.2, etc. Mechanically, the hairline is placed over the number on the C scale, and the reciprocal value of the number is read on the C_1 scale.

The reciprocal scale can also be utilized to simplify multiplication and division. Some examples will demonstrate the theory.

---

sample problem

Multiply $15 \times 60 \times 93$.

---

solution

step 1: Note that the number can be written as

$$\frac{15 \times 60}{\frac{1}{93}}$$

Multiply 15×60. The answer is read as 9 on the D scale.

step 2: Divide the number 9 by $\frac{1}{93}$. Line up 93 of the C_1 scale with 9 of the D scale.

step 3: Move the hairline to the index. Read the solution on the D scale as 837. The decimal value is 83,700.

---

sample problem

$$\text{Evaluate } \frac{64 \times 190}{37}.$$

---

solution

step 1: Divide 64 by 37. Set the hairline to 64 of the D scale and move the C scale until 37 lines up with the hairline.

step 2: Multiply the answer by 190. Since the previous answer is at the index, move the hairline to 190 on the C scale and read the value on the D scale. The number is 3285. The decimal value is 328.5.

---

sample problem

$$\text{Evaluate } \frac{91}{475 \times 320}.$$

---

solution

step 1: Multiply 91 by $\frac{1}{475}$. Set the right index of the C scale over 91 on the D scale.

step 2: Move the hairline to 475 on the C_1 scale.

step 3: Move the C scale until 320 lines up with the hairline.

step 4: Read 598 on the D scale under the index. The decimal value is 5.98×10^{-4}.

student problems

Verify the solutions to the given problems.

1. $130 \times 445 \times 720 = 41.6 \times 10^6$

2. $1.91 \times 27,300 \times 0.00875 = 456$

3. $\dfrac{35,300 \times 17.4}{5250} = 117$

4. $\dfrac{5 \times 111}{23.6 \times 7.35} = 3.2$

5. $\dfrac{3 \times 227}{10 \times 31 \times 89} = 0.0247$

6. $\dfrac{1}{19 \times 179 \times 1360} = 2.16 \times 10^{-6}$

7. $9 \times 2410 \times 1057 = 2.29 \times 10^{7}$

8. $\dfrac{6310 \times 179}{436} = 2.59 \times 10^{3}$

squares and square roots (A and B scales)

Investigation of the A and B scales indicates an index on the left side, and an index in the center, as well as an index on the right side. All values to the left of center will be called A–odd, whereas all values to the right of the center index will be called A–even. The odd or even notation refers to the number of digits in the number for proper placement on the slide rule for evaluation of square roots. Thus, 106, having three digits, is placed on the A–odd side. Consequently, to determine square roots, place the hairline over the given number on the proper A scale. Read the value under the hairline on the D scale. The decimal is evaluated by the power of ten method.

Some illustrative problems will demonstrate the theory.

---

sample problem

Determine the square root of 62.

---

solution

step 1: There are two digits in the number. The number is placed on the A–even scale.

step 2:   Place the hairline over the number 62 on the *A*–even scale.

step 3:   Read the number under the hairline on the *D* scale. The number is 789. The decimal value is 7.89.

---

sample problem

$$\text{Evaluate } \frac{(4.85)^2}{32}.$$

---

solution

step 1:   Place the hairline over 4.85 on the *D* scale.

step 2:   Read the value on the *A*–even scale. The number is 23.5.

step 3:   Place this value on the *D* scale. Move the hairline over to this value.

step 4:   Move the *C* scale so that number 32 lines up with the hairline.

step 5:   Read the answer under the *C* scale index on the *D* scale. The number is 734. The decimal value is 7.34×10^{-1}, or 0.734.

the S, T, and ST scales

The *S* and *T* scales can be used to solve any right triangle if two sides or one side and one acute angle are given. The *S* or sine scale from left to right starts at 5.74° and ends at 90°, corresponding to sine values from approximately 0.1 to 1. The sine of the angle is found by placing the hairline over the angle on the *S* scale and reading the solution under the hairline on the *D* scale.

The *T* or tangent scale from left to right starts at 5.74° and ends at 45°, corresponding to tangent values from approximately 0.1 to 1. The tangent of the angle is found by placing the hairline over the angle on the *T* scale and reading the value under the hairline on the *D* scale. The *T* scale from right to left ranges from 45° to 84.26°, corresponding to tangent values from 1 to 10. These values are read under the hairline on the D_1 scale.

The *ST* scale or sine tangent scale starts at 0.574° and ends at 5.74°. In this region, the sine and tangent are almost equal and correspond to the sines or tangents in the range of 0.01 to 0.1 read on the *D* scale.

On some rulers, the *S* scale has two values corresponding to each major division. One value is in black and refers to the sine of the angle. The other value is in red and refers to the cosine of the angle. Thus, the cosine of the angle goes from right to left on the S scale or from 0 to 84.26°, read directly on the D scale from 1 to 0.1.

Some illustrative problems will demonstrate the theory.

---

sample problem

Determine:

(a) sin 36.9°   (c) sin 26°
(b) sin 50°   (d) sin 18°

---

solution

Make the C and D scales coincide. Then

step 1:   sin 36.9° = 0.6
Set the hairline over 36.9° on the S scale and read the solution under the hairline on the D scale.

step 2:   sin 50° = 0.765
Set the hairline over 50° on the S scale and read the solution under the hairline on the D scale.

step 3:   sin 26° = 0.439
Set the hairline over 26° on the S scale and read the solution under the hairline on the D scale.

step 4:   sin 18° = 0.31
Set the hairline over 18° on the S scale and read the solution under the hairline on the D scale.

---

sample problem

Determine:

(a) tan 20°   (c) tan 65°
(b) tan 38°   (d) tan 43.5°

---

solution

Make the C and D scales coincide.

step 1:   tan 20° = 0.364
Set the hairline over 20° on the T scale and read the solution under the hairline on the D scale.

step 2:   tan 38° = 0.78
> Set the hairline over 38° on the *T* scale and read the solution under the hairline on the *D* scale.

step 3:   tan 65° = 2.15
> Set the hairline over 65° on the *T* scale and read the solution under the hairline on the D_1 scale.

step 4:   tan 43.5°
> Set the hairline over 43.5° on the *T* scale and read the solution under the hairline on the *D* scale.

---

sample problem

Determine:

(a) sin 2.64°	(c) sin 4.3°
(b) tan 0.64°	(d) tan 1.9°

---

solution

step 1:   sin 2.64° = 0.0461
> Set the hairline over 2.64° on the *ST* scale and read the solution under the hairline on the *D* scale.

step 2:   tan 0.64° = 0.01118
> Set the hairline over 0.64° on the *ST* scale and read the solution under the hairline on the *D* scale.

step 3:   sin 4.3° = 0.075
> Set the hairline over 4.3° on the *ST* scale and read the solution under the hairline on the *D* scale.

step 4:   tan 1.9° = 0.0344
> Set the hairline over 1.9° on the *ST* scale and read the solution under the hairline on the *D* scale.

the J operator

Consider the triangle shown in Fig. A–1. Note that *R* is the real component, *X* is the *j* or imaginary component, and *Z* is the hypotenuse of magnitude of the phasor. From trigonometry,

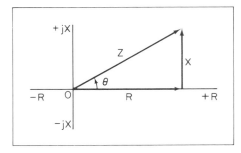

figure A–1
right triangle in imaginary plane

$$\sin \theta = \frac{X}{Z}$$

$$\cos \theta = \frac{R}{Z}$$

$$\tan \theta = \frac{X}{R}$$

$$Z^2 = R^2 + X^2$$

In rectangular form, the complex number is written as $R+jX$, where the j preceding the value indicates that the value is at right angles to any non j terms. Mathematically, $j = \sqrt{-1}$ and is imaginary in nature. Its application in electrical technology is to act as an operator in phasor analysis. Thus,

$$\underbrace{R+jX}_{\substack{\text{rectangular} \\ \text{form}}} = \underbrace{Z\underline{/\theta^\circ}}_{\substack{\text{polar} \\ \text{form}}}$$

In polar notation, the magnitude of the phasor is Z units along an angle of θ°, from the origin.

convert from rectangular to polar form

Since all ac quantities must be treated as phasors, it is necessary for the student to perform calculations in phasor algebra with reasonable accuracy and speed. The proper use of the slide rule will relieve the technician and engineer of the boring and tedious details involved in solving ac phasor problems. Addition and subtraction of phasors must be performed in rectangular form. Multiplication and division are simplified by polar form operation.

The various possibilities that exist in converting $R+jX = Z\underline{/\theta°}$ are

1. R is greater than X.

2. R is ten times or more greater than X.

3. R is 100 times or more greater than X.

4. X is greater than R.

5. X is 10 times or more greater than R.

6. X is 100 times or more greater than R.

7. X is equal to R.

possibility 1

Set the index of the C scale over R on the D scale. Set the hairline to X on the D scale. Read θ at the hairline on the T scale. (Use numbers to the right of major divisions.) Move the slide so that θ on the S scale appears under the hairline. Read Z on the D scale under the index. Sign of θ corresponds to sign of X.

---

sample problem

Convert $5+j3$ to polar form.

---

solution

step 1: Set the index of C over 5 on the D scale.

step 2: Set the hairline to 3 on the D scale.

step 3: Read θ on the T scale. The angle is 30.9°.

step 4: Move the slider portion of the rule so that θ of the S scale lines up with the hairline.

step 5: Read Z on the D scale under the index of the C scale. The value is 5.84.

The answer is $5+j3 = 5.84\underline{/30.9°}$

possibility 2

Set the index of the C scale over R on the D scale. Set the hairline to X on the D scale. Read θ at the hairline on the ST scale. Assume that Z is approximately equal to R. The sign of the angle corresponds to the sign of X.

---

sample problem

Convert $34+j3$ to polar form.

---

solution

step 1:   Set the index of C over 34 on the D scale.

step 2:   Set the hairline to 3 on the D scale.

step 3:   Read θ on the ST scale under the hairline. The angle is 5.05°.

step 4:   The answer is $34+j3 = 34\underline{/+5.05°}$

possibility 3

Assume that θ is approximately equal to zero and that the impedance Z is approximately equal to R.

example

$$300+j1 = 300\underline{/0°}$$
$$4250+j24 = 4250\underline{/0°}$$

possibility 4

Set the right index of the C scale at X on the D scale. Set the hairline over R on the D scale. Read θ under the hairline on the T scale. (Use numbers to the left of the major divisions.) Move the slide so that θ on the S scale appears under the hairline. Use the numbers to the left of the major divisions. Read Z on the D scale under the index of the C scale. The sign of θ corresponds to the sign of X.

---

sample problem

Convert $3+j5$ to polar form.

---

solution

step 1:　Set the index of the C scale at 5 on the D scale.

step 2:　Set the hairline over 3 on the D scale.

step 3:　Read θ under the hairline on the T scale. The angle is 59.1°.

step 4:　Move the slide so that the angle 59.1° on the S scale appears under the hairline. (Use numbers to the left of the major divisions.)

step 5:　Read the value of Z on the D scale under the index of C. The answer is $5.84\underline{/59.1°}$.

possibility 5

Set the index of the C scale over X on the D scale. Set the hairline over R on the D scale. Read the angle θ under the hairline on the ST scale. Compute the angle by subtracting this value from 90°. Z is approximately equal to X.

---

sample problem

Convert $3+j34$ to polar form.

---

solution

step 1:　Set the index of C over 34 on the D scale.

step 2:　Set the hairline over 3 on the D scale.

step 3:　Read the angle on the ST scale and subtract this angle from 90°. The answer is 84.95°.

step 4:　Z is approximately equal to 34. The answer is $34\underline{/84.95°}$.

possibility 6

The angle is approximately equal to 90°. The sign of the angle corresponds to the sign of X. The impedance Z is approximately equal to X.

example

$$1+\ j300 = \ 300\underline{/90°}$$
$$24+j4250 = 4250\underline{/90°}$$

In this case the angle is 45°, the sign depending on X. Z is equal to 1.414 multiplied by R or X, since they are both equal.

convert from polar to rectangular form

Consider the triangle used as reference which was shown in Fig. A–1. The problem that exists is: Given the value Z/θ, determine the value of R and X. From the trigonometric ratios given, it is evident that

$$R = Z \cos \theta$$

and

$$X = Z \sin \theta$$

Thus, the conversion is

$$Z/\theta = Z(\cos \theta + j \sin \theta) = R + jX$$

The mechanics of slide rule usage are as follows: Set the index of the C scale over Z on the D scale. Move the hairline to angle θ on the S scale. The sine of the angle is read from left to right. The cosine of the angle is read from right to left. Usually, if the slide rule has colors, the black numbers define the sine and the red numbers the cosine. After the angle is determined, read the result on the D scale under the hairline.

---

sample problem

Convert $14/40°$ to rectangular form.

---

solution

step 1: Place the index of the C scale over 14 on the D scale.

step 2: Move the hairline to black 40 on the S scale.

step 3: Read the answer on the D scale under the hairline. The result is 9.00. This is the j term.

step 4: Since the red 40 is off the rule, place the other index over 14 on the D scale.

step 5:   Move the hairline over to the red 40 on the S scale.

step 6:   Read the answer on the D scale under the hairline. The result is 10.72.

The total answer is $10.72 + j9$.

In many problems, the phasor does not lie in the first quadrant. It should be noted that the following trigonometric identities can be used.

$$\sin (180° \pm \theta) = \sin \pm \theta$$
and
$$\cos (180° \pm \theta) = \cos \theta$$

Also, note that the trigonometric functions are all positive in the first quadrant, the sine is positive in the second quadrant, the tangent is positive in the third quadrant, and the cosine is positive in the fourth quadrant.

Two sets of 20 problems each follow. The second set contains answers to the first set, and vice versa.

Problems	*Answers*
1. $68 + j56$	**1.** $88.1\underline{/39.5°}$
2. $81 - j12$	**2.** $81.9\underline{/-8.42°}$
3. $918 - j237$	**3.** $948\underline{/-14.5°}$
4. $24 + j18$	**4.** $30\underline{/36.9°}$
5. $8.55 - j1.96$	**5.** $8.78\underline{/-12.9°}$
6. $18.32 - j7.64$	**6.** $19.85\underline{/-22.6°}$
7. $63.5 + j110$	**7.** $127\underline{/60°}$
8. $12.7 - j71.6$	**8.** $72.6\underline{/-79.9°}$
9. $28 - j28$	**9.** $41\underline{/-45°}$
10. $70.7 + j70.7$	**10.** $100\underline{/+45°}$
11. $7.63 + j13.2$	**11.** $15.3\underline{/60°}$
12. $34.6 - j20$	**12.** $40\underline{/-30°}$
13. $11 - j1.9$	**13.** $11.2\underline{/-9.8°}$

14. $47.3 + j925$

14. $925\underline{/87°}$

15. $4 + j16$

15. $16.5\underline{/76°}$

16. $43.1 + j17.2$

16. $46.4\underline{/21.8°}$

17. $5.5 + j9.52$

17. $11\underline{/60°}$

18. $38.3 + j32.2$

18. $50\underline{/40°}$

19. $67.6 + j27.4$

19. $73\underline{/22°}$

20. $27 - j560$

20. $560\underline{/-87.3°}$

cubes and cube roots using the K scale

The K scale and the D scale are the two scales used in finding cubes and cube roots. Note that the K scale has two unit ones between the left and right index. Any number on the K scale has its cube root on the D scale. Any number on the D scale has its cube on the K scale. Split the K scale into thirds as follows: The left index to the first index will be K_L, denoting K left; between the two middle indices will be K_m, denoting K middle; the other third is K_R, denoting K right. For numbers between one and 10, use K_L; for numbers between 10 and 100, use K_m; for numbers between 100 and 1000, use K_R.

To determine which portion of the K scale to use, the power of ten technique is used. Move the decimal point in groups of three until there are three or less digits to the left of the decimal point. An example will demonstrate the technique.

---

sample problem 1

Evaluate the cube root of 4,680,000.

---

solution

Move the decimal point in groups of three according to the rule, and the result is

$$\sqrt[3]{4.68 \times 10^6}$$

Consequently, K_L must be used. The solution is

$$\sqrt[3]{4.68 \times 10^6} = 1.67 \times 10^2 = 167$$

sample problem 2

Evaluate the cube root of 0.00000081

solution

Move the decimal point in groups of three, obtaining

$$\sqrt[3]{810 \times 10^{-9}}$$

K_R must be used. The answer is

$$\sqrt[3]{810 \times 10^{-9}} = 9.34 \times 10^{-3}$$

sample problem 3

Evaluate the cube power of 43.2

solution

Move the hairline over 432 on the D scale and read 81 on the K scale. Using powers of ten, we find that the solution is

$$(4.32 \times 10)^3 = 81 \times 10^3$$

the L scale

The L scale is the common logarithm scale. The theory of logarithms requires an understanding of the *characteristic* and *mantissa*. A positive characteristic is equal to $(n-1)$, where n is equal to the number of digits counted to the left of the decimal.

A *positive mantissa* is the power of ten required to yield a number from 1 to 10. A *negative mantissa* is the power of 10 required to yield a number from 0.1 to 1. It is the mantissa that is found on the L scale. The base ten is the basis of the L scale and does not appear on the rule.

An example will demonstrate the technique.

sample problem

Evaluate log 3800.

solution

The positive characteristic is $(4-1)$ equals 3. To determine the mantissa, move the hairline over 38 on the D scale and read 0.58 on the L scale. The correct answer is $\log 3800 = 3.58$.

---

sample problem

Evaluate $\log 0.005$.

---

solution

Rewrite the problem as $\log 5 + \log 10^{-3}$. The characteristic of the $\log 5$ is zero, and the characteristic of the $\log 10^{-3}$ is -3. To determine the mantissa, move the hairline over 5 on the D scale and read 0.699 on the L scale. Thus, $\log 0.005 = 0.699 - 3.0000 = -2.301$.

---

sample problem

Evaluate $10^{0.3}$

---

solution

Let $x = 10^{0.3}$.

$\log x = 0.3 \log 10 = 0.3$

Taking antilogarithms of both sides results in

$x = \log^{-1} 0.3$

Move the hairline of the rule over 0.3 on the L scale and read the answer of 2 on the D scale. The characteristic of 0 denotes the power of 10. Thus, the correct solution is

$10^{0.3} = 2$

---

sample problem

Evaluate $10^{2.8} = x$

---

solution

$\log x = 2.8$

$x = \log^{-1} 2.8 = 630$

the LL scales

The proper operation of the *LL* and *D* scales solves the problem of evaluating the natural log base, symbolized by epsilon ($\epsilon = 2.718$) to a power and vice versa. The terminology used to denote the natural logarithm is ln, which is a logarithm to the base epsilon.

Note the values of the *LL*1, *LL*2, and *LL*3, respectively. The range of values for *LL*1 is from 1.01 to 1.105. *LL*2 continues from 1.105 to approximately 2.718. *LL*3 continues from 2.718 to about 22,000.

Consequently, it is evident that the three scales range from 1.01 to 22,000 for natural logarithm usage. To raise epsilon to powers of 0.01 to 0.1, use the *LL*1 scale. The *LL*2 scale raises epsilon to a power of 0.1 to 1; the *LL*3 scale is used to raise epsilon to a power ranging from 1 to 10. Thus, to tabulate the results, let $e^x = y$. Then

table A–1

scale	x	y
*LL*1	0.01–0.1	1.01–1.105
*LL*2	0.1–1	1.105–2.718
*LL*3	1–10	2.718–22,000
	D scale	*LL* scales

Some examples will demonstrate the technique of solving natural logarithmic problems.

sample problem

Evaluate x in the equation $\epsilon^x = 20.1$.

solution

Take the natural logarithm of both sides. Thus,

$$x = \ln 20.1$$

Move the hairline over 20.1 on the *LL*3 scale and read the number 3 on the *D* scale.

sample problem

$$\text{Evaluate } \epsilon^2 = y.$$

solution

Move the hairline over 2 on the D scale and read the answer on the $LL3$ scale equal to 7.4.

B
table of exponentials

table 1   exponentials

					value of ϵ^n					
n	0	1	2	3	4	5	6	7	8	9
0.0	1.000	1.010	1.020	1.030	1.041	1.051	1.062	1.073	1.083	1.094
0.1	1.105	1.116	1.127	1.139	1.150	1.162	1.174	1.185	1.197	1.209
0.2	1.221	1.234	1.246	1.259	1.271	1.284	1.297	1.310	1.323	1.336
0.3	1.350	1.363	1.377	1.391	1.405	1.419	1.433	1.448	1.462	1.477
0.4	1.492	1.507	1.522	1.537	1.553	1.568	1.584	1.600	1.616	1.632
0.5	1.649	1.665	1.682	1.699	1.716	1.733	1.751	1.768	1.786	1.804
0.6	1.822	1.840	1.859	1.878	1.896	1.916	1.935	1.954	1.974	1.994
0.7	2.014	2.034	2.054	2.075	2.096	2.117	2.138	2.160	2.181	2.203
0.8	2.226	2.248	2.270	2.293	2.316	2.340	2.363	2.387	2.411	2.435
0.9	2.460	2.484	2.509	2.535	2.560	2.586	2.612	2.638	2.664	2.691
1.0	2.718	2.746	2.773	2.801	2.829	2.858	2.886	2.915	2.945	2.974
.1	3.004	3.034	3.065	3.096	3.127	3.158	3.190	3.222	3.254	3.287
.2	3.320	3.353	3.387	3.421	3.456	3.490	3.525	3.561	3.597	3.633
.3	3.669	3.706	3.743	3.781	3.819	3.857	3.896	3.935	3.975	4.015
.4	4.055	4.096	4.137	4.179	4.221	4.263	4.306	4.349	4.393	4.437
.5	4.482	4.527	4.572	4.618	4.665	4.711	4.759	4.807	4.855	4.904
.6	4.953	5.003	5.053	5.104	5.155	5.207	5.259	5.312	5.366	5.419
.7	5.474	5.529	5.585	5.641	5.697	5.755	5.812	5.871	5.930	5.989
.8	6.050	6.110	6.172	6.234	6.297	6.360	6.424	6.488	6.554	6.619
.9	6.686	6.753	6.821	6.890	6.959	7.029	7.099	7.171	7.243	7.316
2.0	7.389	7.463	7.538	7.614	7.691	7.768	7.846	7.925	8.004	8.085
.1	8.166	8.248	8.331	8.415	8.499	8.585	8.671	8.758	8.846	8.935
.2	9.025	9.116	9.207	9.300	9.393	9.488	9.583	9.679	9.777	9.875
.3	9.974	10.07	10.18	10.28	10.38	10.49	10.59	10.70	10.81	10.91
.4	11.02	11.13	11.25	11.36	11.47	11.59	11.71	11.82	11.94	12.06
.5	12.18	12.31	12.43	12.55	12.68	12.81	12.94	13.07	13.20	13.33
.6	13.46	13.60	13.74	13.87	14.01	14.15	14.30	14.44	14.59	14.73
.7	14.88	15.03	15.18	15.33	15.49	15.64	15.80	15.96	16.12	16.28
.8	16.45	16.61	16.78	16.95	17.12	17.29	17.46	17.64	17.81	17.99
.9	18.17	18.36	18.54	18.73	18.92	19.11	19.30	19.49	19.69	19.89
3.0	20.09	20.29	20.49	20.70	20.91	21.12	21.33	21.54	21.76	21.98
.1	22.20	22.42	22.65	22.87	23.10	23.34	23.57	23.81	24.05	24.29
.2	24.53	24.78	25.03	25.28	25.53	25.79	26.05	26.31	26.58	26.84
.3	27.11	27.39	27.66	27.94	28.22	28.50	28.79	29.08	29.37	29.67
.4	29.96	30.27	30.57	30.88	31.19	31.50	31.82	32.14	32.46	32.79
.5	33.11	33.45	33.78	34.12	34.47	34.81	35.16	35.52	35.87	36.23
.6	36.60	36.97	37.34	37.71	38.09	38.47	38.86	39.25	39.65	40.05
.7	40.45	40.85	41.26	41.68	42.10	42.52	42.95	43.38	43.82	44.26
.8	44.70	45.15	45.60	46.06	46.53	46.99	47.47	47.94	48.42	48.91
.9	49.40	49.90	50.40	50.91	51.42	51.93	52.46	52.99	53.52	54.05

table 1 (contd.)   exponentials

					value of ϵ^{-n}					
$-n$	0	1	2	3	4	5	6	7	8	9
0.0	1.0000	0.9900	0.9802	0.9975	0.9608	0.9512	0.9418	0.9324	0.9231	0.9139
.1	.9048	.8958	.8869	.8781	.8694	.8607	.8521	.8437	.8353	.8270
.2	.8187	.8106	.8025	.7945	.7866	.7788	.7711	.7634	.7558	.7483
.3	.7408	.7334	.7262	.7189	.7118	.7047	.6977	.6907	.6839	.6771
.4	.6703	.6637	.6571	.6505	.6440	.6376	.6313	.6250	.6188	.6126
.5	.6065	.6005	.5945	.5886	.5828	.5770	.5712	.5655	.5599	.5543
.6	.5488	.5434	.5379	.5326	.5273	.5221	.5169	.5117	.5066	.5016
.7	.4966	.4916	.4868	.4819	.4771	.4724	.4677	.4630	.4584	.4538
.8	.4493	.4449	.4404	.4361	.4317	.4274	.4232	.4190	.4148	.4107
.9	.4066	.4025	.3985	.3946	.3906	.3867	.3829	.3791	.3753	.3716
1.0	.3679	.3642	.3606	.3570	.3535	.3499	.3465	.3430	.3396	.3362
.1	.3329	.3296	.3263	.3230	.3198	.3166	.3135	.3104	.3073	.3042
.2	.3012	.2982	.2952	.2923	.2894	.2865	.2837	.2808	.2780	.2753
.3	.2725	.2698	.2671	.2645	.2619	.2592	.2567	.2541	.2516	.2491
.4	.2466	.2441	.2417	.2393	.2369	.2346	.2322	.2299	.2276	.2254
.5	.2231	.2209	.2187	.2165	.2144	.2123	.2001	.2081	.2060	.2039
.6	.2019	.1999	.1979	.1959	.1940	.1921	.1901	.1883	.1864	.1845
.7	.1827	.1809	.1791	.1773	.1755	.1738	.1720	.1703	.1686	.1670
.8	.1653	.1637	.1620	.1604	.1588	.1572	.1557	.1541	.1526	.1511
.9	.1496	.1481	.1466	.1451	.1437	.1423	.1409	.1395	.1381	.1367
2.0	.1353	.1340	.1327	.1313	.1300	.1287	.1275	.1262	.1249	.1237
.1	.1225	.1212	.1200	.1188	.1177	.1165	.1153	.1142	.1130	.1119
.2	.1108	.1097	.1086	.1075	.1065	.1054	.1043	.1033	.1023	.1013
.3	.1003	.0993	.0983	.0973	.0963	.0954	.0944	.0935	.0925	.0916
.4	.0907	.0898	.0889	.0880	.0872	.0863	.0854	.0846	.0837	.0829
.5	.0821	.0813	.0805	.0797	.0789	.0781	.0773	.0765	.0758	.0750
.6	.0743	.0735	.0728	.0721	.0714	.0707	.0699	.0693	.0686	.0679
.7	.0672	.0665	.0659	.0652	.0646	.0639	.0633	.0627	.0620	.0614
.8	.0608	.0602	.0596	.0590	.0584	.0578	.0563	.0567	.0551	.0556
.9	.0550	.0545	.0539	.0534	.0529	.0523	.0518	.0513	.0508	.0503
3.0	.0498	.0493	.0488	.0483	.0478	.0474	.0469	.0464	.0460	.0455
.1	.0450	.0446	.0442	.0437	.0433	.0429	.0424	.0420	.0416	.0412
.2	.0408	.0404	.0400	.0396	.0392	.0388	.0384	.0380	.0376	.0373
.3	.0369	.0365	.0361	.0358	.0354	.0351	.0347	.0344	.0340	.0337
.4	.0334	.0330	.0327	.0324	.0321	.0317	.0314	.0311	.0308	.0305
.5	.0302	.0299	.0296	.0293	.0290	.0287	.0284	.0281	.0279	.0276
.6	.0273	.0271	.0268	.0265	.0263	.0260	.0257	.0255	.0252	.0250
.7	.0247	.0245	.0242	.0240	.0238	.0235	.0233	.0231	.0228	.0226
.8	.0224	.0221	.0219	.0217	.0215	.0213	.0211	.0209	.0207	.0204
.9	.0202	.0200	.0198	.0196	.0194	.0193	.0191	.0189	.0187	.0185

C
natural trigonometric functions

sine

angles from 0°0 to 44°9

	.0	.1	.2	.3	.4	.5	.6	.7	.8	.9
0°	.0000	.0017	.0035	.0052	.0070	.0087	.0105	.0122	.0140	.0157
1°	.0175	.0192	.0209	.0227	.0244	.0262	.0279	.0297	.0314	.0332
2°	.0349	.0366	.0384	.0401	.0419	.0436	.0454	.0471	.0488	.0506
3°	.0523	.0541	.0558	.0576	.0593	.0610	.0628	.0645	.0663	.0680
4°	.0698	.0715	.0732	.0750	.0767	.0785	.0802	.0819	.0837	.0854
5°	.0872	.0889	.0906	.0924	.0941	.0958	.0976	.0993	.1011	.1028
6°	.1045	.1063	.1080	.1097	.1115	.1132	.1149	.1167	.1184	.1201
7°	.1219	.1236	.1253	.1271	.1288	.1305	.1323	.1340	.1357	.1374
8°	.1392	.1409	.1426	.1444	.1461	.1478	.1495	.1513	.1530	.1547
9°	.1564	.1582	.1599	.1616	.1633	.1650	.1668	.1685	.1702	.1719
10°	.1736	.1754	.1771	.1788	.1805	.1822	.1840	.1857	.1874	.1891
11°	.1908	.1925	.1942	.1959	.1977	.1994	.2011	.2028	.2045	.2062
12°	.2079	.2096	.2113	.2130	.2147	.2164	.2181	.2198	.2215	.2233
13°	.2250	.2267	.2284	.2300	.2317	.2334	.2351	.2368	.2385	.2402
14°	.2419	.2436	.2453	.2470	.2487	.2504	.2521	.2538	.2554	.2571
15°	.2588	.2605	.2622	.2639	.2656	.2672	.2689	.2706	.2723	.2740
16°	.2756	.2773	.2790	.2807	.2823	.2840	.2857	.2874	.2890	.2907
17°	.2924	.2940	.2957	.2974	.2990	.3007	.3024	.3040	.3057	.3074
18°	.3090	.3107	.3123	.3140	.3156	.3173	.3190	.3206	.3223	.3239
19°	.3256	.3272	.3289	.3305	.3322	.3338	.3355	.3371	.3387	.3404
20°	.3420	.3437	.3453	.3469	.3486	.3502	.3518	.3535	.3551	.3567
21°	.3584	.3600	.3616	.3633	.3649	.3665	.3681	.3697	.3714	.3730
22°	.3746	.3762	.3778	.3795	.3811	.3827	.3843	.3859	.3875	.3891
23°	.3907	.3923	.3939	.3955	.3971	.3987	.4003	.4019	.4035	.4051
24°	.4067	.4083	.4099	.4115	.4131	.4147	.4163	.4179	.4195	.4210
25°	.4226	.4242	.4258	.4274	.4289	.4305	.4321	.4337	.4352	.4368
26°	.4384	.4399	.4415	.4431	.4446	.4462	.4478	.4493	.4509	.4524
27°	.4540	.4555	.4571	.4586	.4602	.4617	.4633	.4648	.4664	.4679
28°	.4695	.4710	.4726	.4741	.4756	.4772	.4787	.4802	.4818	.4833
29°	.4848	.4863	.4879	.4894	.4909	.4924	.4939	.4955	.4970	.4985
30°	.5000	.5015	.5030	.5045	.5060	.5075	.5090	.5105	.5120	.5135
31°	.5150	.5165	.5180	.5195	.5210	.5225	.5240	.5255	.5270	.5284
32°	.5299	.5314	.5329	.5344	.5358	.5373	.5388	.5402	.5417	.5432
33°	.5446	.5461	.5476	.5490	.5505	.5519	.5534	.5548	.5563	.5577
34°	.5592	.5606	.5621	.5635	.5650	.5664	.5678	.5693	.5707	.5721
35°	.5736	.5750	.5764	.5779	.5793	.5807	.5821	.5835	.5850	.5864
36°	.5878	.5892	.5906	.5920	.5934	.5948	.5962	.5976	.5990	.6004
37°	.6018	.6032	.6046	.6060	.6074	.6088	.6101	.6115	.6129	.6143
38°	.6157	.6170	.6184	.6198	.6211	.6225	.6239	.6252	.6266	.6280
39°	.6293	.6307	.6320	.6334	.6347	.6361	.6374	.6388	.6401	.6414
40°	.6428	.6441	.6455	.6468	.6481	.6494	.6508	.6521	.6534	.6547
41°	.6561	.6574	.6587	.6600	.6613	.6626	.6639	.6652	.6665	.6678
42°	.6691	.6704	.6717	.6730	.6743	.6756	.6769	.6782	.6794	.6807
43°	.6820	.6833	.6845	.6858	.6871	.6884	.6896	.6909	.6921	.6934
44°	.6947	.6959	.6972	.6984	.6997	.7009	.7022	.7034	.7046	.7059

angles from 45°0 to 89°9

	.0	.1	.2	.3	.4	.5	.6	.7	.8	.9
45°	.7071	.7083	.7096	.7108	.7120	.7133	.7145	.7157	.7169	.7181
46°	.7193	.7206	.7218	.7230	.7242	.7254	.7266	.7278	.7290	.7302
47°	.7314	.7325	.7337	.7349	.7361	.7373	.7385	.7396	.7408	.7420
48°	.7431	.7443	.7455	.7466	.7478	.7490	.7501	.7513	.7524	.7536
49°	.7547	.7559	.7570	.7581	.7593	.7604	.7615	.7627	.7638	.7649
50°	.7660	.7672	.7683	.7694	.7705	.7716	.7727	.7738	.7749	.7760
51°	.7771	.7782	.7793	.7804	.7815	.7826	.7837	.7848	.7859	.7869
52°	.7880	.7891	.7902	.7912	.7923	.7934	.7944	.7955	.7965	.7976
53°	.7986	.7997	.8007	.8018	.8028	.8039	.8049	.8059	.8070	.8080
54°	.8090	.8100	.8111	.8121	.8131	.8141	.8151	.8161	.8171	.8181
55°	.8192	.8202	.8211	.8221	.8231	.8241	.8251	.8261	.8271	.8281
56°	.8290	.8300	.8310	.8320	.8329	.8339	.8348	.8358	.8368	.8377
57°	.8387	.8396	.8406	.8415	.8425	.8434	.8443	.8453	.8462	.8471
58°	.8480	.8490	.8499	.8508	.8517	.8526	.8536	.8545	.8554	.8563
59°	.8572	.8581	.8590	.8599	.8607	.8616	.8625	.8634	.8643	.8652
60°	.8660	.8669	.8678	.8686	.8695	.8704	.8712	.8721	.8729	.8738
61°	.8746	.8755	.8763	.8771	.8780	.8788	.8796	.8805	.8813	.8821
62°	.8829	.8838	.8846	.8854	.8862	.8870	.8878	.8886	.8894	.8902
63°	.8910	.8918	.8926	.8934	.8942	.8949	.8957	.8965	.8973	.8980
64°	.8988	.8996	.9003	.9011	.9018	.9026	.9033	.9041	.9048	.9056
65°	.9063	.9070	.9078	.9085	.9092	.9100	.9107	.9114	.9121	.9128
66°	.9135	.9143	.9150	.9157	.9164	.9171	.9178	.9184	.9191	.9198
67°	.9205	.9212	.9219	.9225	.9232	.9239	.9245	.9252	.9259	.9265
68°	.9272	.9278	.9285	.9291	.9298	.9304	.9311	.9317	.9323	.9330
69°	.9336	.9342	.9348	.9354	.9361	.9367	.9373	.9379	.9385	.9391
70°	.9397	.9403	.9409	.9415	.9421	.9426	.9432	.9438	.9444	.9449
71°	.9455	.9461	.9466	.9472	.9478	.9483	.9489	.9494	.9500	.9505
72°	.9511	.9516	.9521	.9527	.9532	.9537	.9542	.9548	.9553	.9558
73°	.9563	.9568	.9573	.9578	.9583	.9588	.9593	.9598	.9603	.9608
74°	.9613	.9617	.9622	.9627	.9632	.9636	.9641	.9646	.9650	.9655
75°	.9659	.9664	.9668	.9673	.9677	.9681	.9686	.9690	.9694	.9699
76°	.9703	.9707	.9711	.9715	.9720	.9724	.9728	.9732	.9736	.9740
77°	.9744	.9748	.9751	.9755	.9759	.9763	.9767	.9770	.9774	.9778
78°	.9781	.9785	.9789	.9792	.9796	.9799	.9803	.9806	.9810	.9813
79°	.9816	.9820	.9823	.9826	.9829	.9833	.9836	.9839	.9842	.9845
80°	.9848	.9851	.9854	.9857	.9860	.9863	.9866	.9869	.9871	.9874
81°	.9877	.9880	.9882	.9885	.9888	.9890	.9893	.9895	.9898	.9900
82°	.9903	.9905	.9907	.9910	.9912	.9914	.9917	.9919	.9921	.9923
83°	.9925	.9928	.9930	.9932	.9934	.9936	.9938	.9940	.9942	.9943
84°	.9945	.9947	.9949	.9951	.9952	.9954	.9956	.9957	.9959	.9960
85°	.9962	.9963	.9965	.9966	.9968	.9969	.9971	.9972	.9973	.9974
86°	.9976	.9977	.9978	.9979	.9980	.9981	.9982	.9983	.9984	.9985
87°	.9986	.9987	.9988	.9989	.9990	.9990	.9991	.9992	.9993	.9993
88°	.9994	.9995	.9995	.9996	.9996	.9997	.9997	.9997	.9998	.9998
89°	.9998	.9999	.9999	.9999	.9999	1.000	1.000	1.000	1.000	1.000

cosine

angles from 0°0 to 44°9

	.0	.1	.2	.3	.4	.5	.6	.7	.8	.9
0°	1.0000	1.0000	1.0000	1.0000	1.0000	1.0000	.9999	.9999	.9999	.9999
1°	.9998	.9998	.9998	.9997	.9997	.9997	.9996	.9996	.9995	.9995
2°	.9994	.9993	.9993	.9992	.9991	.9990	.9990	.9989	.9988	.9987
3°	.9986	.9985	.9984	.9983	.9982	.9981	.9980	.9979	.9978	.9977
4°	.9976	.9974	.9973	.9972	.9971	.9969	.9968	.9966	.9965	.9963
5°	.9962	.9960	.9959	.9957	.9956	.9954	.9952	.9951	.9949	.9947
6°	.9945	.9943	.9942	.9940	.9938	.9936	.9934	.9932	.9930	.9928
7°	.9925	.9923	.9921	.9919	.9917	.9914	.9912	.9910	.9907	.9905
8°	.9903	.9900	.9898	.9895	.9893	.9890	.9888	.9885	.9882	.9880
9°	.9877	.9874	.9871	.9869	.9866	.9863	.9860	.9857	.9854	.9851
10°	.9848	.9845	.9842	.9839	.9836	.9833	.9829	.9826	.9823	.9820
11°	.9816	.9813	.9810	.9806	.9803	.9799	.9796	.9792	.9789	.9785
12°	.9781	.9778	.9774	.9770	.9767	.9763	.9759	.9755	.9751	.9748
13°	.9744	.9740	.9736	.9732	.9728	.9724	.9720	.9715	.9711	.9707
14°	.9703	.9699	.9694	.9690	.9686	.9681	.9677	.9673	.9668	.9664
15°	.9659	.9655	.9650	.9646	.9641	.9636	.9632	.9627	.9622	.9617
16°	.9613	.9608	.9603	.9598	.9593	.9588	.9583	.9578	.9573	.9568
17°	.9563	.9558	.9553	.9548	.9542	.9537	.9532	.9527	.9521	.9516
18°	.9511	.9505	.9500	.9494	.9489	.9483	.9478	.9472	.9466	.9461
19°	.9455	.9449	.9444	.9438	.9432	.9426	.9421	.9415	.9409	.9403
20°	.9397	.9391	.9385	.9379	.9373	.9367	.9361	.9354	.9348	.9342
21°	.9336	.9330	.9323	.9317	.9311	.9304	.9298	.9291	.9285	.9278
22°	.9272	.9265	.9259	.9252	.9245	.9239	.9232	.9225	.9219	.9212
23°	.9205	.9198	.9191	.9184	.9178	.9171	.9164	.9157	.9150	.9143
24°	.9135	.9128	.9121	.9114	.9107	.9100	.9092	.9085	.9078	.9070
25°	.9063	.9056	.9048	.9041	.9033	.9026	.9018	.9011	.9003	.8996
26°	.8988	.8980	.8973	.8965	.8957	.8949	.8942	.8934	.8926	.8918
27°	.8910	.8902	.8894	.8886	.8878	.8870	.8862	.8854	.8846	.8838
28°	.8829	.8821	.8813	.8805	.8796	.8788	.8780	.8771	.8763	.8755
29°	.8746	.8738	.8729	.8721	.8712	.8704	.8695	.8686	.8678	.8669
30°	.8660	.8652	.8643	.8634	.8625	.8616	.8607	.8599	.8590	.8581
31°	.8572	.8563	.8554	.8545	.8536	.8526	.8517	.8508	.8499	.8490
32°	.8480	.8471	.8462	.8453	.8443	.8434	.8425	.8415	.8406	.8396
33°	.8387	.8377	.8368	.8358	.8348	.8339	.8329	.8320	.8310	.8300
34°	.8290	.8281	.8271	.8261	.8251	.8241	.8231	.8221	.8211	.8202
35°	.8192	.8181	.8171	.8161	.8151	.8141	.8131	.8121	.8111	.8100
36°	.8090	.8080	.8070	.8059	.8049	.8039	.8028	.8018	.8007	.7997
37°	.7986	.7976	.7965	.7955	.7944	.7934	.7923	.7912	.7902	.7891
38°	.7880	.7869	.7859	.7848	.7837	.7826	.7815	.7804	.7793	.7782
39°	.7771	.7760	.7749	.7738	.7727	.7716	.7705	.7694	.7683	.7672
40°	.7660	.7649	.7638	.7627	.7615	.7604	.7593	.7581	.7570	.7559
41°	.7547	.7536	.7524	.7513	.7501	.7490	.7478	.7466	.7455	.7443
42°	.7431	.7420	.7408	.7396	.7385	.7373	.7361	.7349	.7337	.7325
43°	.7314	.7302	.7290	.7278	.7266	.7254	.7242	.7230	.7218	.7206
44°	.7193	.7181	.7169	.7157	.7145	.7133	.7120	.7108	.7096	.7083

angles from 45°0 to 89°9

	.0	.1	.2	.3	.4	.5	.6	.7	.8	.9
45°	.7071	.7059	.7046	.7034	.7022	.7009	.6997	.6984	.6972	.6959
46°	.6947	.6934	.6921	.6909	.6896	.6884	.6871	.6858	.6845	.6833
47°	.6820	.6807	.6794	.6782	.6769	.6756	.6743	.6730	.6717	.6704
48°	.6691	.6678	.6665	.6652	.6639	.6626	.6613	.6600	.6587	.6574
49°	.6561	.6547	.6534	.6521	.6508	.6494	.6481	.6468	.6455	.6441
50°	.6428	.6414	.6401	.6388	.6374	.6361	.6347	.6334	.6320	.6307
51°	.6293	.6280	.6266	.6252	.6239	.6225	.6211	.6198	.6184	.6170
52°	.6157	.6143	.6129	.6115	.6101	.6088	.6074	.6060	.6046	.6032
53°	.6018	.6004	.5990	.5976	.5962	.5948	.5934	.5920	.5906	.5892
54°	.5878	.5864	.5850	.5835	.5821	.5807	.5793	.5779	.5764	.5750
55°	.5736	.5721	.5707	.5693	.5678	.5664	.5650	.5635	.5621	.5606
56°	.5592	.5577	.5563	.5548	.5534	.5519	.5505	.5490	.5476	.5461
57°	.5446	.5432	.5417	.5402	.5388	.5373	.5358	.5344	.5329	.5314
58°	.5299	.5284	.5270	.5255	.5240	.5225	.5210	.5195	.5180	.5165
59°	.5150	.5135	.5120	.5105	.5090	.5075	.5060	.5045	.5030	.5015
60°	.5000	.4984	.4970	.4955	.4939	.4924	.4909	.4894	.4879	.4863
61°	.4848	.4833	.4818	.4802	.4787	.4772	.4756	.4741	.4726	.4710
62°	.4695	.4679	.4664	.4648	.4633	.4617	.4602	.4586	.4571	.4555
63°	.4540	.4524	.4509	.4493	.4478	.4462	.4446	.4431	.4415	.4399
64°	.4384	.4368	.4352	.4337	.4321	.4305	.4289	.4274	.4258	.4242
65°	.4226	.4210	.4195	.4179	.4163	.4147	.4131	.4115	.4099	.4083
66°	.4067	.4051	.4035	.4019	.4003	.3987	.3971	.3955	.3939	.3923
67°	.3907	.3891	.3875	.3859	.3843	.3827	.3811	.3795	.3778	.3762
68°	.3746	.3730	.3714	.3697	.3681	.3665	.3649	.3633	.3616	.3600
69°	.3584	.3567	.3551	.3535	.3518	.3502	.3486	.3469	.3453	.3437
70°	.3420	.3404	.3387	.3371	.3355	.3338	.3322	.3305	.3289	.3272
71°	.3256	.3239	.3223	.3206	.3190	.3173	.3156	.3140	.3123	.3107
72°	.3090	.3074	.3057	.3040	.3024	.3007	.2990	.2974	.2957	.2940
73°	.2924	.2907	.2890	.2874	.2857	.2840	.2823	.2807	.2790	.2773
74°	.2756	.2740	.2723	.2706	.2689	.2672	.2656	.2639	.2622	.2605
75°	.2588	.2571	.2554	.2538	.2521	.2504	.2487	.2470	.2453	.2436
76°	.2419	.2402	.2385	.2368	.2351	.2334	.2317	.2300	.2284	.2267
77°	.2250	.2233	.2215	.2198	.2181	.2164	.2147	.2130	.2113	.2096
78°	.2079	.2062	.2045	.2028	.2011	.1994	.1977	.1959	.1942	.1925
79°	.1908	.1891	.1874	.1857	.1840	.1822	.1805	.1788	.1771	.1754
80°	.1736	.1719	.1702	.1685	.1668	.1650	.1633	.1616	.1599	.1582
81°	.1564	.1547	.1530	.1513	.1495	.1478	.1461	.1444	.1426	.1409
82°	.1392	.1374	.1357	.1340	.1323	.1305	.1288	.1271	.1253	.1236
83°	.1219	.1201	.1184	.1167	.1149	.1132	.1115	.1097	.1080	.1063
84°	.1045	.1028	.1011	.0993	.0976	.0958	.0941	.0924	.0906	.0889
85°	.0872	.0854	.0837	.0819	.0802	.0785	.0767	.0750	.0732	.0715
86°	.0698	.0680	.0663	.0645	.0628	.0610	.0593	.0576	.0558	.0541
87°	.0523	.0506	.0488	.0471	.0454	.0436	.0419	.0401	.0384	.0366
88°	.0349	.0332	.0314	.0297	.0279	.0262	.0244	.0227	.0209	.0192
89°	.0175	.0157	.0140	.0122	.0105	.0087	.0070	.0052	.0035	.0017

tangent

angles from 0°0 to 44°9

	.0	.1	.2	.3	.4	.5	.6	.7	.8	.9
0°	.0000	.0017	.0035	.0052	.0070	.0087	.0105	.0122	.0140	.0157
1°	.0175	.0192	.0209	.0227	.0244	.0262	.0279	.0297	.0314	.0332
2°	.0349	.0367	.0384	.0402	.0419	.0437	.0454	.0472	.0489	.0507
3°	.0524	.0542	.0559	.0577	.0594	.0612	.0629	.0647	.0664	.0682
4°	.0699	.0717	.0734	.0752	.0769	.0787	.0805	.0822	.0840	.0857
5°	.0875	.0892	.0910	.0928	.0945	.0963	.0981	.0998	.1016	.1033
6°	.1051	.1069	.1086	.1104	.1122	.1139	.1157	.1175	.1192	.1210
7°	.1228	.1246	.1263	.1281	.1299	.1317	.1334	.1352	.1370	.1388
8°	.1405	.1423	.1441	.1459	.1477	.1495	.1512	.1530	.1548	.1566
9°	.1584	.1602	.1620	.1638	.1655	.1673	.1691	.1709	.1727	.1745
10°	.1763	.1781	.1799	.1817	.1835	.1853	.1871	.1890	.1908	.1926
11°	.1944	.1962	.1980	.1998	.2016	.2035	.2053	.2071	.2089	.2107
12°	.2126	.2144	.2162	.2180	.2199	.2217	.2235	.2254	.2272	.2290
13°	.2309	.2327	.2345	.2364	.2382	.2401	.2419	.2438	.2456	.2475
14°	.2493	.2512	.2530	.2549	.2568	.2586	.2605	.2623	.2642	.2661
15°	.2679	.2698	.2717	.2736	.2754	.2773	.2792	.2811	.2830	.2849
16°	.2867	.2886	.2905	.2924	.2943	.2962	.2981	.3000	.3019	.3038
17°	.3057	.3076	.3096	.3115	.3134	.3153	.3172	.3191	.3211	.3230
18°	.3249	.3269	.3288	.3307	.3327	.3346	.3365	.3385	.3404	.3424
19°	.3443	.3463	.3482	.3502	.3522	.3541	.3561	.3581	.3600	.3620
20°	.3640	.3659	.3679	.3699	.3719	.3739	.3759	.3779	.3799	.3819
21°	.3839	.3859	.3879	.3899	.3919	.3939	.3959	.3979	.4000	.4020
22°	.4040	.4061	.4081	.4101	.4122	.4142	.4163	.4183	.4204	.4224
23°	.4245	.4265	.4286	.4307	.4327	.4348	.4369	.4390	.4411	.4431
24°	.4452	.4473	.4494	.4515	.4536	.4557	.4578	.4599	.4621	.4642
25°	.4663	.4684	.4706	.4727	.4748	.4770	.4791	.4813	.4834	.4856
26°	.4877	.4899	.4921	.4942	.4964	.4986	.5008	.5029	.5051	.5073
27°	.5095	.5117	.5139	.5161	.5184	.5206	.5228	.5250	.5272	.5295
28°	.5317	.5340	.5362	.5384	.5407	.5430	.5452	.5475	.5498	.5520
29°	.5543	.5566	.5589	.5612	.5635	.5658	.5681	.5704	.5727	.5750
30°	.5774	.5797	.5820	.5844	.5867	.5890	.5914	.5938	.5961	.5985
31°	.6009	.6032	.6056	.6080	.6104	.6128	.6152	.6176	.6200	.6224
32°	.6249	.6273	.6297	.6322	.6346	.6371	.6395	.6420	.6445	.6469
33°	.6494	.6519	.6544	.6569	.6594	.6619	.6644	.6669	.6694	.6720
34°	.6745	.6771	.6796	.6822	.6847	.6873	.6899	.6924	.6950	.6976
35°	.7002	.7028	.7054	.7080	.7107	.7133	.7159	.7186	.7212	.7239
36°	.7265	.7292	.7319	.7346	.7373	.7400	.7427	.7454	.7481	.7508
37°	.7536	.7563	.7590	.7618	.7646	.7673	.7701	.7729	.7757	.7785
38°	.7813	.7841	.7869	.7898	.7926	.7954	.7983	.8012	.8040	.8069
39°	.8098	.8127	.8156	.8185	.8214	.8243	.8273	.8302	.8332	.8361
40°	.8391	.8421	.8451	.8481	.8511	.8541	.8571	.8601	.8632	.8662
41°	.8693	.8724	.8754	.8785	.8816	.8847	.8878	.8910	.8941	.8972
42°	.9004	.9036	.9067	.9099	.9131	.9163	.9195	.9228	.9260	.9293
43°	.9325	.9358	.9391	.9424	.9457	.9490	.9523	.9556	.9590	.9623
44°	.9657	.9691	.9725	.9759	.9793	.9827	.9861	.9896	.9930	.9965

angles from 45°0 to 89°9

	.0	.1	.2	.3	.4	.5	.6	.7	.8	.9
45°	1.000	1.003	1.007	1.011	1.014	1.018	1.021	1.025	1.028	1.032
46°	1.036	1.039	1.043	1.046	1.050	1.054	1.057	1.061	1.065	1.069
47°	1.072	1.076	1.080	1.084	1.087	1.091	1.095	1.099	1.103	1.107
48°	1.111	1.115	1.118	1.122	1.126	1.130	1.134	1.138	1.142	1.146
49°	1.150	1.154	1.159	1.163	1.167	1.171	1.175	1.179	1.183	1.188
50°	1.192	1.196	1.200	1.205	1.209	1.213	1.217	1.222	1.226	1.230
51°	1.235	1.239	1.244	1.248	1.253	1.257	1.262	1.266	1.271	1.275
52°	1.280	1.285	1.289	1.294	1.299	1.303	1.308	1.313	1.317	1.322
53°	1.327	1.332	1.337	1.342	1.347	1.351	1.356	1.361	1.366	1.371
54°	1.376	1.381	1.387	1.392	1.397	1.402	1.407	1.412	1.418	1.423
55°	1.428	1.433	1.439	1.444	1.450	1.455	1.460	1.466	1.471	1.477
56°	1.483	1.488	1.494	1.499	1.505	1.511	1.517	1.522	1.528	1.534
57°	1.540	1.546	1.552	1.558	1.564	1.570	1.576	1.582	1.588	1.594
58°	1.600	1.607	1.613	1.619	1.625	1.632	1.638	1.645	1.651	1.658
59°	1.664	1.671	1.678	1.684	1.691	1.698	1.704	1.711	1.718	1.725
60°	1.732	1.739	1.746	1.753	1.760	1.767	1.775	1.782	1.789	1.797
61°	1.804	1.811	1.819	1.827	1.834	1.842	1.849	1.857	1.865	1.873
62°	1.881	1.889	1.897	1.905	1.913	1.921	1.929	1.937	1.946	1.954
63°	1.963	1.971	1.980	1.988	1.997	2.006	2.014	2.023	2.032	2.041
64°	2.050	2.059	2.069	2.078	2.087	2.097	2.106	2.116	2.125	2.135
65°	2.145	2.154	2.164	2.174	2.184	2.194	2.204	2.215	2.225	2.236
66°	2.246	2.257	2.267	2.278	2.289	2.300	2.311	2.322	2.333	2.344
67°	2.356	2.367	2.379	2.391	2.402	2.414	2.426	2.438	2.450	2.463
68°	2.475	2.488	2.500	2.513	2.526	2.539	2.552	2.565	2.578	2.592
69°	2.605	2.619	2.633	2.646	2.660	2.675	2.689	2.703	2.718	2.733
70°	2.747	2.762	2.778	2.793	2.808	2.824	2.840	2.856	2.872	2.888
71°	2.904	2.921	2.937	2.954	2.971	2.989	3.006	3.024	3.042	3.060
72°	3.078	3.096	3.115	3.133	3.152	3.172	3.191	3.211	3.230	3.251
73°	3.271	3.291	3.312	3.333	3.354	3.376	3.398	3.420	3.442	3.465
74°	3.487	3.511	3.534	3.558	3.582	3.606	3.630	3.655	3.681	3.706
75°	3.732	3.758	3.785	3.812	3.839	3.867	3.895	3.923	3.952	3.981
76°	4.011	4.041	4.071	4.102	4.134	4.165	4.198	4.230	4.264	4.297
77°	4.331	4.366	4.402	4.437	4.474	4.511	4.548	4.586	4.625	4.665
78°	4.705	4.745	4.787	4.829	4.872	4.915	4.959	5.005	5.050	5.097
79°	5.145	5.193	5.242	5.292	5.343	5.396	5.449	5.503	5.558	5.614
80°	5.671	5.730	5.789	5.850	5.912	5.976	6.041	6.107	6.174	6.243
81°	6.314	6.386	6.460	6.535	6.612	6.691	6.772	6.855	6.940	7.026
82°	7.115	7.207	7.300	7.396	7.495	7.596	7.700	7.806	7.916	8.028
83°	8.144	8.264	8.386	8.513	8.643	8.777	8.915	9.058	9.205	9.357
84°	9.514	9.677	9.845	10.02	10.20	10.39	10.58	10.78	10.99	11.20
85°	11.43	11.66	11.91	12.16	12.43	12.71	13.00	13.30	13.62	13.95
86°	14.30	14.67	15.06	15.46	15.89	16.35	16.83	17.34	17.89	18.46
87°	19.08	19.74	20.45	21.20	22.02	22.90	23.86	24.90	26.03	27.27
88°	28.64	30.14	31.82	33.69	35.80	38.19	40.92	44.07	47.74	52.08
89°	57.29	63.66	71.62	81.85	95.49	114.6	143.2	191.0	286.5	573.0

index

A

A, (symbol for ammeter)
ac, 190
ac generator, 183
Acceptor atom, 28
Addition complex, 218
  by phasor methods, 218
Admittance:
  series and parallel, 262
  parameters, 333
  polar form, 267
Air core coils, 247
Air gaps, 155
Algebra complex, 218
Alternating current, 183
  average value, 196
  effective value, 192
  instantaneous value, 184
  peak, 184
  sine wave, 184
Alternating current circuits, 191
  impedance networks, 238
  parallel, 258
  power in, 193
  series, 238
  simple, 199
American wire gauge, 17
Ammeter, 61
  ac, 61
  dc, 61
  multirange, 63
  symbol, 61
Ammeter shunts, 62
Ampere, definition of, 7
Ampere turn, 144
Amplitude of sine wave, 184
Analysis:
  dimensional, 132
  loop, 41
  nodal, 50
Angle:
  impedance, 240
  lag, 201
  lead, 203
  power factor, 199
  phase, 187
  radians, 185

Angular velocity, 186
Antiresonant frequency, 296
Apparent power, 243
Atom:
  acceptor, 28
  donor, 25
  structure, 5
Average permeability, 145
Average value of:
  current, 196
  power, 199
  voltage, 196

B

B, (symbol for magnetic flux density)
B, (symbol for susceptance)
BH curves, 148
Balanced bridge, 72
Bandwidth of resonant circuit, 293, 301
Bridge, wheatstone, 72

C

C, (symbol for capacitance)
CR time constant, 132
Capacitance, 120
  ac circuits, 202
  dc circuits, 129
  definition of, 121
  factors governing, 122
  farad, 121
Capacitive reactance:
  definition, 203
  factors governing, 203
Capacitors:
  ceramic, 127
  charging, 129
  construction, 120
  definition, 119
  dielectrics, 121
  discharging, 134
  electrolytic, 127
  energy stored, 137
  integrated circuit, 128
  mica, 126
  paper, 126
  parallel, 123
  series, 123

Capacitors: (*contd*)
  symbol, 120
  variable, 128
Charge:
  electric, 118
  in capacitors, 118
  carriers, 25
  coulomb, 7
  definition, 7
  negative, 25
  positive, 25
Charging current, 131
Circuit:
  ac, 190
  CR, 129, 132
  coupled, 345
  dc, 37
  electric, 30
  equivalent, 328
  magnetic, 143
  magnetically coupled, 157, 345
  parallel, 295
  resonant, 288
  series, 288
  series parallel, 305
Circular mil, 15
Coefficient of coupling, 344
Coefficient of temperature, 18
Coincident symmetry, 369
Color code, 21
Complex waves, 362
Complex algebra, 211
Conductance, 259
Conductivity, 13
Conductors, electric, 10
  semi, 22
Conjugate quantity, 223
Constant voltage source, 40
Conventional current direction, 8
Conversion factors, 4
Coordinates:
  polar, 214
  rectangular, 213
Coulomb's law, 7
Coupled circuits, 344
Coupling:
  coefficient, 344
  mutual, 344
Coupling networks:
  four terminal, 345
  hybrid parameters, 334
  mutual impedance, 344
  open circuit impedance parameters, 332
  short circuit admittance parameters, 333
Current:
  alternating, 183
  ampere, 7
  branch, 46
  capacitive, 131
  charging, 131
  collapse, 134
  conventional, 8
  definition of, 6
  dependent variable, 332
  direct, 37
  divider theorem, 49
  electric, 7

Current: (*contd*)
  electron flow, 8
  inductive, 170
  instantaneous, 184
  lag, 235
  lead, 237
  line, 240
  loop, 41
  magnetizing, 143
  mesh, 41
  nodal, 99
  nonsinusoidal, 194
  phase, 203
  rate of change, 166
  reactive, 203
  resonant, 290
  rise, 131
  short circuit, 91
  sine wave, 184
  source, 183
  tank, 300
Current carriers, 22
Current law, (see Kirchhoff's laws)

D

dc, (see direct coupling)
D'Arsonval movement, 59
D'Arsonval construction, 60
Decay of current, 134
Degrees, mechanical and electrical, 185
Delta–Wye connection, 76
Determinants, 109
Diagram:
  admittance, 99
  impedance, 252
  phasor, 214
  symbols, 29
Dielectric constant, 121
  table of, 121
Diffused silicon resistors, 26
Direct current, 41
Discharge:
  capacitor, 134
  inductor, 175
Divider:
  current, 49
  voltage, 42
Donor atoms, 25
Driving point impedance, 346

E

E, (symbol for emf)
emf, (see electromotive force)
Effective value of:
  current, 191
  nonsinusoidal wave, 194
  sine wave, 194
  voltage, 196
Efficiency, 33
Electric charge, 119
Electric circuit, (see circuit)
Electric current, (see current)
Electric energy, 34
Electric field, 6, 119
Electromagnetic induction, 164
Electrons:
  charge, 6

Electrons: (*contd*)
  drift, 6
  energy levels, 5
  flow, 7
  free, 25
  mass, 6
  shells, 7
  theory, 6
  velocity, 7
Energy:
  definition of, 34
  electrical, 35
  electron, 35
  heat form, 23
  in capacitor, 137
  in inductor, 174
  levels in an atom, 5
  shells, 5
Equations:
  general bridge, 72
  general transformer, 348
  Kirchhoff's law, 41
  loop, 41
  mesh, 41
  nodal, 99
Equivalent circuits:
  circuit simplifications, 328
  delta – wye, 76
  Norton's, 91, 322
  parallel, 46
  series, 41
  Thevenin's, 88, 320
  transformer, 350
Exponential curves, 132, 170

F

f, (symbol for frequency)
F, (symbol for electromagnetic force)
Factor:
  conversion, 4
  power, 243
  quality, 290
Farad, definition, 121
Field:
  electric, 119
  magnetic, 144
Figure of merit, 290
Flux, magnetic, 144
  definition, 145
  density, 120
  leakage, 342
  lines of force, 144
  linkage, 344
  mutual, 344
  weber, 143
Flux density, 145
Flux magnetic, 144
Force:
  electromotive, 183
  magnetic, 144
Forward current ratio, 334
Four terminal network, 311
Fourier series, 362
Free electrons, 25
Frequency:
  definition, 185
  fundamental, 362

Frequency (*contd*)
  harmonic, 362
  resonant, 289

G

G, (symbol for conductance)
Galvonometer, 72
Gauge wire, 17
Gauss (flux density), 144
General bridge equation, 72
General transformer equations, 348
Generation of ac, 183
Geometrical construction vector diagram, 218
Graphical determination of rms value, 191
graphic symbols, 29

H

H, (symbol for magnetic field intensity)
Half cycle:
  average, 197
  effective value, 194
Half power point frequencies, 293
Harmonics:
  definition of, 363
  evaluation of, 364
Henry:
  definition of, 167
  unit of permeance, 145
Hole flow, 23
Hybrid parameters, 334
Hysteresis:
  magnetic, 147
  loop, 148

I

I, (symbol for electric current)
IR drop, 10
Imaginary component, 218
Impedance:
  conjugate, 222
  coupled, 342
  definition of, 239
  input, 346
  mutual, 341
  open circuit, 332
  parallel, 259
  reflected, 341
  series, 239, 269
  series parallel, 281
  short circuit, 333
Impedance angle, 240
Impedance diagram, 239
Impedance matching, 323
Independent variable of, 332
Inductance:
  ac circuits, 201
  dc circuits, 170
  factors governing, 167
  henry, 167
  mutual, 341
  self, 166
Induction:
  electromagnetic, 164
  magnetic, 164
  mutual, 347
  self, 166

Inductive reactance:
    definition of, 201
    factors governing, 201
Inductors:
    definition of, 167
    energy stored, 174
    graphic symbol, 167
    parallel, 169
    series, 168
Instantaneous value:
    current, 184
    power, 197
    voltage, 184
Instruments:
    D'Arsonval movement, 59
Insulators, 10
Intensity:
    electric field, 119
    magnetic field, 144
Iron, magnetic properties of, 148

J

j operator, 211
joule, 9
junction, 22

K

Kirchhoff's law equations, 41, 46
Kirchhoff's laws:
    ac circuits, 311, 314
    dc circuits, 41
    current law, 46
    voltage law, 41

L

L, (symbol for inductance)
Laws,
    Coulomb's, 7
    Faraday's, 164
    Kirchhoff's, 41
    Lenz's, 166
    Ohm's, 30
Leakage flux, 342
Lenz's law, 166
Lines of force:
    electric, 34
    magnetic, 144
Loop:
    closed, 97
    current, 97
    equations, 311

M

M, (symbol for mutual inductance)
MKS units, 1
mmf, (see magnetomotive force)
Magnetic circuits, 143
    parallel, 157
    series, 149
Magnetic retentivity, 148
Magnetism:
    residual, 148
Magnetization, 143
Magnetization curves, 149
Magnetizing current, 144

Magnetomotive force, 144
Maximum power transfer, 323
Meter movements:
    D'Arsonval, 59
Meters:
    ammeters, 61
    current, 61
    ohmmeter, 70
    voltage, 64
Meter calibrating resistor, 65
Meter multiplier resistor, 65
Meter shunt resistor, 62
Mhos, 259
Millman's theorem, 92
Mirror symmetry, 367
Multimeter, 64
Multiplier voltmeter, 64
Mutual coupling, 347
Mutual flux, 342
Mutual inductance, 347

N

N type semiconductor, 25
Networks:
    bridge, 72
    coupling, 347
    delta, 76
    four terminal, 331
    pi, 76
    resistance, 13
    T, 76
    wheatstone, 72
Network theorems:
    maximum power transfer, 95
    Millman's, 92
    Norton's, 91
    reciprocity, 93
    superposition, 82
    Thevenin's, 88
Nodal equations, 99, 314
Nonsinusoidal waves:
    definition, 362
    effective value, 195
    harmonics, 363
    power of, 377
Norton's theorem, 91
Notation:
    double subscript, 328
Nucleus, 5

O

Ohm, 10
Ohmmeter, 70
Ohm's law, 30
Open circuit impedance parameters, 332
Operator j, 211
Orbits electron, 5
Output admittance, 333

P

P, (symbol for power)
P type semiconductor, 24
Pn junction, 22
Parallel:
    capacitors, 123
    inductors, 169
    resistors, 258

Parallel circuits:
  electric, 258
  magnetic, 157
Parallel impedances, 259
Parallel resonance, 295
Parameters:
  admittance, 333
  hybrid, 334
  impedance, 332
Peak value of sine wave, 184
Period, 184
Permeability, 145
Phase angle, 187
Phasor:
  addition, 218
  components, 213
  diagram, 227
  division, 222
  exponential representation of, 215
  multiplication by, 221
  polar form, 214
  rectangular form, 213
  reference, 242
  rotation of, 227
  subtraction of, 220
  trigonometric form, 213
Pi network, 76
Point symmetry, 368
Polar coordinates, 214
Polarity of coils, 347
Potential difference, 8
Potentiometer, 29
Power:
  apparent, 243
  average, 200
  definition, 32
  factor, 243
  instantaneous, 204
  maximum transfer of, 323
  measurement, 32
  reactive, 247
  true, 243
  watt, 247
Power factor:
  angle, 243
  definition, 243
  lagging, 250
  leading, 245
  unity, 298

Q

Q, (symbol for electric charge)
Q, (symbol for quality factor)
Q of parallel resonant circuit, 297
  of series resonant circuit, 290

R

R, (symbol for resistance)
Radian, 185
Reactance:
  capacitive, 203
  inductive, 201
  mutual, 349
  net, 252
Reactive power, 247
Rectangular coordinates, 213

Reference axis, 238
Reluctance, 144
Resistance:
  ac, 199
  color code, 21
  definition, 13
  equivalent, 38
  factors governing, 13
  nature of, 13
  ohm, 10
  temperature coefficient, 18
  total, 38
Resistance network, 37
Resistivity, 13
Resistors:
  construction, 13
  definition, 13
  graphic symbol, 21
  meter calibrating, 65
  multiplier, 65
  parallel, 62
  semiconductor, 22
  series, 65
  series dropping, 65
  shunt meter, 62
Resonance:
  definition, 289
  multiple, 305
  parallel, 295
  series, 288
Resonant frequency, 289
Response of networks, 311
Response curves, 292
Reverse voltage ratio, 334
Right hand rule, 143
Rise current:
  CR circuit, 132
  LR circuit, 170
Rotating magnetic field, 183
Rotating phasor, 211

S

Saturation magnetic, 148
Sawtooth wave, 369
Schematic diagram, 29
Secondary of transformer, 347
Selectivity of resonant circuit, 291
Self inductance:
  definition, 167
  factors governing, 167
  graphic symbols, 167
Semiconductors, 22
Sensitivity of resonant circuit, 291
Sensitivity of voltmeter, 68
Series capacitors, 123
Series circuit:
  electric, 30
  magnetic, 143
Series dropping resistor, 65
Series Fourier, 362
Series impedance, 239
Series inductors, 168
Series resonance, 289
Sharpness of resonance, 291
Sheet resistance, 26
Shells, electron, 5
Short circuit admittance parameters, 333

Short circuit current, 91, 322
Shunt ammeter, 62
Sine wave:
    addition, 197
    amplitude, 184
    average value, 196
    cycle, 184
    definition of, 184
    effective value, 192
    instantaneous value, 184
    peak value, 184
    period, 185
    rms value, 192
Source:
    of emf, 29
    constant current, 29
    constant voltage, 29
Square mil, 15
Superposition theorem, 82
Susceptance, 263
Symbols, graphic, 29
Symmetrical waveforms, 367
Symmetry, 366

T

t, (symbol for time)
T, (symbol for temperature)
T network, 76, 325, 328
Tables:
    color code for resistors, 21
    conversion units, 2
    dielectric constants, 121
    energy conversion factors, 2
    exponentials, (see Appendix)
    graphic symbols, 29
    interrelationship of network parameters,
        334
    powers of ten, 3
    specific resistances, 14
    temperature coefficients, 14
    trigonometric functions, (see Appendix)
    wire gage, 17
Tank current, 300
Temperature:
    coefficient of resistance, 19
    inferred zero resistance, 18
Theorems:
    maximum power transfer, 323
    Millman's, 92
    Norton's, 91, 322
    superposition, 82, 318
    Thevenin's, 88, 320
Thevenin's theorem, 88
Time constant:
    CR network, 132
    LR network, 170
Transformation:
    impedance, 328
    pi to tee, 76
    tee to pi, 76
Transformer:
    air core, 347
    audio, 354
    equivalent circuit, 350
    testing, 355
Transformer equations, 348
Transformer tests, 355

Transient current:
    in CR circuit, 131
    in LR circuit, 171
True power, 247
Turns ratio, 345

U

Units:
    conversion, 2
    interrelationship of, 2
    MKS, 1
Universal exponential graphs:
    CR circuits, 132
    LR circuits, 170

V

V, (symbol for voltage drop)
Valence electrons, 5
Vector:
    addition, 218
        by rectangular coordinates, 218
    division, 222
    multiplication, 221
    rotating, 227
    subtracting, 227
Vector algebra, 213
Voltage:
    definition, 9
    divider, 42
    generated, 183
    induced, 343
    open circuit, 88
    phase, 234
    reactive, 245
Voltmeters:
    sensitivity, 68
    symbol, 29

W

W, (symbol for work and energy)
Watt, 32
Wave:
    nonsinusoidal, 363
    sawtooth, 365
    sine, 190
    square, 371
    symmetry, 366
Weber, 143
Wheatstone bridge, 72
Wire table, 17
Work, 9
Wye–delta transformation, 76

X

X, (symbol for reactance)

Y

Y, (symbol for admittance)
y parameters, 333
Y systems:
    current in, 99
    voltage in, 99

Z

Z, (symbol for impedance)
z parameters, 332
zero axis symmetry, 368